Minerals and Lipids Profiles in Cardiovascular Disorders in South Asia

Nayab Batool Rizvi • Saeed Ahmad Nagra

Minerals and Lipids Profiles in Cardiovascular Disorders in South Asia

Cu, Mg, Se, Zn and Lipid Serum Profiles for the Example of Patients in Pakistan

Springer

Nayab Batool Rizvi
Saeed Ahmad Nagra
New Campus, Institute of Chemistry
University of the Punjab
Lahore
Pakistan

ISBN 978-3-642-34248-6 ISBN 978-3-642-34249-3 (eBook)
DOI 10.1007/978-3-642-34249-3
Springer Heidelberg New York Dordrecht London

Printed on acid-free paper

Springer is part of Springer Science+Business Media (www.springer.com)

Preface

Cardiovascular disorders are on the rise in developing countries, and SARC countries are no exception to it. Data on any aspect of these disorders, with reference to local living conditions, is very scarce. The work presented here is part of a doctoral study, and the findings primarily focus on the aim of providing an insight about the prevalence of some of the serum minerals and lipids profiles present in various cardiovascular disorders. The data presented and discussed in this book pertains to Pakistan but can serve as a base line, or reference, for future studies in the country, as well as the surrounding continent. It is expected that data presented here may pave the way to more advanced research regarding cardiovascular problems for researchers/clinicians wishing to proceed in this direction.

The linguistics of the book has been kept very simple for easy comprehension so that researchers, clinicians, and even students may be void of any problem. Detailed tables and figures have also been included, which provide, at a glance, a true idea of the prevalence of serum minerals and lipids profiles in countries such as Pakistan. Our findings have been presented as observed, and it is hoped that the book will prove to be of great importance to the habitants of South Asian countries.

We hope that the users of this book will find it appealing and reliable. Any suggestions for improvement will be most welcome.

Lahore, 2012

Nayab Batool Rizvi
Saeed Ahmad Nagra

Contents

Chapter 1
Introduction: Minerals

1.1 Introduction

Cardiovascular diseases (CVD) are a group of disorders of the heart and blood vessels and include coronary heart disease (CHD), cerebrovascular disease, raised blood pressure (hypertension), rheumatic heart disease, angina, peripheral arterial disease, atherosclerosis, congenital heart disease, heart failure, and deep vein thrombosis (WHO 2009). Cardiovascular disease is the leading cause of death worldwide accounting for about 40 % of global mortality and 10.3 % of the global burden of disease. The incidence of CVD has risen greatly in low- and middle-income countries (Yusuf et al. 2001). Among cardiovascular disease, CHD is the most common heart disease, responsible for 40–60 % of all deaths and is invariably associated with atherosclerosis (thickening of the arteries). Atherosclerosis is a major cause of CVD that develops slowly over decades. In atherosclerosis, narrowing of the vessel's lumen may lead to obstruction and to its clinical manifestations, such as myocardial infarction (MI) (Lefant and Savage 1995). Although women do not suffer from this problem as frequently as men, but it is numerically more significant in the case of older women. It has been reported that the risk of CHD is more in postmenopausal women (Miller 1990). Angina pectoris is one of the several expressions and one of possible manifestations of coronary heart disease. It is precipitated when the oxygen demands of the myocardium exceeds the available arterial oxygen supply. The major modifiable risk factors for CVD include high blood pressure (hypertension), high total cholesterol, obesity, physical inactivity, unhealthy diet, diabetes mellitus (DM), and homocysteine levels (Mackay and Mensah 2004). The more risk factors a person has, the greater the likelihood of developing CVD. Risk of CHD is two- to threefold higher in persons having type II diabetes as compared to general population (Fox et al. 2004). In diabetes, cardiovascular disease affects both men and women to the same extent (Uusitupa et al. 1993). It has been observed that women with diabetes, regardless of menopausal status, have a four- to sixfold higher risk of developing CVD, whereas men with diabetes have a two- to threefold increased risk of CVD compared to

N.B. Rizvi and S.A. Nagra, *Minerals and Lipids Profiles in Cardiovascular Disorders in South Asia*, DOI 10.1007/978-3-642-34249-3_1,

women and men without diabetes (Legato et al. 2006). Individuals having DM also face the risk of developing atherosclerosis due to the disturbances in the lipid metabolism (Adlerberth et al. 1998). Furthermore, high-density lipoprotein cholesterol (HDL cholesterol) that is commonly regarded as protective is decreased in diabetic patients. Diabetes mellitus acts as an independent risk factor for several forms of CVD, and it has been estimated that approximately 65 % of CVD-caused deaths are of the persons with DM (Geiss et al. 1995). Diabetes accounts for 6 % of total global mortality, with 50 % of diabetes-associated deaths being attributed to CVD (International Diabetes Federation 2008).

Hypertension is another important risk factor and the main cause of mortality for cardiovascular disease (Whelton 1994). It is reported that 50 % of cases of CVD were caused by elevated blood pressure. Women with CHD are at two- to threefold increased risk of developing hypertension (Wenger 2003). Hypertension affects 8–14 % of the population worldwide, and its prevalence is especially high in diabetic persons (Howard et al. 1996). Hypertension depends on factors such as obesity, physical inactivity, excessive alcohol consumption, sodium and potassium intake, and psychological stress.

Lipids and lipoproteins are well-known risk factors for CHD. Increased levels of triglyceride (TG), total cholesterol (TC), LDL cholesterol, and decreased levels of HDL cholesterol are documented as risk factors for atherogenesis as HDL cholesterol is associated with cholesterol removal from peripheral tissues. Excessive caloric intake, with high component of fat, is associated with an increased serum level of total cholesterol (TC) and LDL cholesterol (Shekelle et al. 1981) and hence increases the risk of CVD. The role of various minerals, in relation to various risk factors such as DM, hypertension, and MI, is discussed in the subsequent lines.

Human health depends on a delicate balance among reactions within the organism in which nerve, muscle, blood, bone, endocrine, and visceral tissues are continually renewed. Vital exchanges that constantly occur involve many enzymatic systems activated by minerals or trace elements (Speich et al. 2001). Most of the minerals in our diet come directly from plants or indirectly from animal sources. A well-balanced, low-fat, antioxidant vitamin-rich diet is one of the important elements of secondary prevention of CVD (Waśkiewicz et al. 2008). Nutrition is strongly associated with socioeconomic conditions, health status, and functional capacity. A wide range of trace elements has been linked with one or another aspect of CVD, some being regarded as beneficial to the heart and blood vessels and others directly or indirectly harmful (Shaper et al. 1979).

1.2 Copper

Copper (Cu) may play a role in cardiovascular disease through its involvement in the coagulation cascade (Linder and Hazegh-Azam 1996). Heart, brain, kidney, and skeletal muscles are considered as important sites where copper is present in substantial concentrations. Only a small fraction of copper occurs in the free form. In the body copper shifts between the cuprous (Cu^{1+}) and cupric (Cu^{2+}) forms, though the majority of the body's copper is in the cupric form. Three nonspecific mechanisms of damage implicated in cardiovascular defects of copper deficiency include peroxidation, glycation, and nitration (Aliabadi 2008). Copper deficiency has been linked to MI, severe tachycardia, and also sudden death due to rupture of the heart. Study shows that Cu deficiency can occur in the presence of homocysteine, especially elevated levels of homocysteine. Homocysteine chelates copper, decreasing its availability to cells (Linnebank et al. 2006). Cardiovascular patients with copper deficiency show several characteristics such as abnormal electrocardiograms, decreased myocardial copper, glucose intolerance, hypercholesterolemia, hyperuricemia, necrosis of myocardial cells, and sudden death. Copper deficiency may also lead to abnormal vessel wall formation and other pathological changes, which may lead to cardiovascular disorders (Vlad et al. 1993).

1.3 Magnesium

It is estimated that magnesium (Mg) is needed for more than 300 biochemical reactions in the body. It helps maintain normal muscle and nerve function, keeps heart rhythm steady, supports a healthy immune system, and keeps bones strong. It also helps regulate blood sugar levels, promotes normal blood pressure, and is known to be involved in energy metabolism, protein synthesis, and phosphorylation reactions, such as muscle insulin tyrosine kinase (Suarez et al. 1995). Of all the cardiovascular risk factors, Mg now takes first place as judged by the accumulation of epidemiological, pathophysiological, clinical, and experimental data. It has been reported that there is an inverse association between dietary magnesium intake and incidence of CVD (Al-Delaimy et al. 2004). It is also reported that increased oxidative stress during magnesium deficiency leads to protein peroxidation, in the early stages and progression of atherosclerotic lesions (Mazur et al. 2007). Magnesium is needed for the electrical stability of the myocardium and prevention of irregular arrhythmias by regulating the flux of cellular potassium levels across cell membrane and transmembrane potentials (Vitale 1992).

It has also been estimated that magnesium may reduce the risk of CHD as a result of inhibiting platelet function, smooth muscle contraction, and by reducing free fatty acids in circulation (Teragawa et al. 2000). Oral magnesium supplementation lowers systolic and diastolic blood pressure and thus prevents the development of hypertension (Kh et al. 2000). Studies have indicated the effects of magnesium deficiency in the hormonal system which controls blood pressure. Magnesium deficiency might affect blood pressure values, leading to hypertension. Various studies also show an inverse association between magnesium (serum and dietary) and blood pressure values (Ma et al. 2006).

1.4 Selenium

One of the major roles of selenium along with zinc and copper in the body is to act as a cofactor of key antioxidant enzymes, namely superoxide dismutase (SOD) and glutathione peroxidase (GSH-Px) and selenoproteins (Nayak et al. 2001). Selenium (Se), an essential component of the antioxidant enzyme GSH-Px, functions as an antioxidant scavenging H_2O_2 and by reducing lipid hydroperoxides to their subsequent less reactive end products. It is an essential trace mineral involved in protection against oxidative damage via selenium-dependent GSH-Px and other selenoproteins. Selenomethionine and selenocysteine are two selenium-containing amino acids that have been detected in proteins (Rayman 2000). The major selenoproteins in plasma is selenoproteins P (SelP). SelP provides more than 50 % of the total plasma selenium (Mostert 2000). Selenium modulates the cellular response to oxidative stress, inducing a faster restoration of the endogenous antioxidative defense system against the production of reactive oxygen species. Selenium deficiency has been implicated in the development of CVD. It has been observed that low Se levels have a casual effect in the development and deterioration of CHD, but it is unknown whether this is due to Se-mediated reduction in lipid peroxidation, inhibition of inflammation, or an improved lipid profile in the blood (Ravn-Haren et al. 2008). By shifting prostaglandin synthesis from prostacyclin to thromboxane, low selenium may increase platelet aggregability and vasoconstriction. Adequate selenium intake therefore helps to maintain adequate nitric oxide concentrations and reduce LDL-cholesterol oxidation. Selenium may also protect the cardiovascular system from toxic metals that have been implicated in atherogenesis, such as mercury, cadmium, and arsenic, by preventing metal-induced oxidative damage or by forming inactive complexes with metals (Feroci et al. 2005). An inverse correlation has been reported between the appearance of some cardiopathies and low Se levels in the environment, diet, and blood (Navarro-Alarcon and López-Martínez 2000). High concentrations of serum selenium predict reduced levels of oxidative stress and subclinical cyclooxygenase-mediated inflammation. Therefore, the association between selenium, oxidative stress, and inflammation may be related to the cardiovascular protective properties of selenium (Helmersson et al. 2005). Measurement of selenium and plasma homocysteine concentrations in elderly humans showed that blood Se levels should be considered as a potential factor to lower total plasma homocysteine, and there exist an independent inverse association between serum Se and plasma homocysteine (Gonzalez et al. 2004). Selenium deficiency has also been associated with a higher incidence of MI and increased mortality rates from CVD. Some studies in diabetics suggest that selenium supplementation may help to prevent vascular complications and that diabetic patients may be deficient in selenium relative to healthy persons (Rajpathak et al. 2005).

1.5 Zinc

Zinc (Zn) is another micronutrient with known antioxidant activity. Zinc is an important component of biomembranes and an essential cofactor in a variety of enzymes (Powell 2000). Zinc plays an important role in the synthesis and function of insulin and is capable of modulating insulin action, and it improves hepatic binding of insulin. It has been observed that people with diabetes have lower serum levels of zinc. Study shows that low serum Zn levels are an independent risk factor for CHD mortality. Studies have also shown that serum zinc level is lower in diabetic patients than in nondiabetic subjects due to increased urinary zinc excretion (Chausmer 1998). Zinc supplementation shows antioxidant properties in the case of diabetics (Roussel et al. 2003) and hence decreases lipid peroxidation. Low serum/plasma zinc concentrations in people with established atherosclerosis indicate that low zinc levels are associated with atherosclerosis (Lissa et al. 2006). Zn supplementation reduces accumulation of cholesterol in the aorta and reduces a number of markers of cholesterol and lipid oxidation (Jenner et al. 2007).

References

Adlerberth AM, Rosengren A, Wilhelmsen L (1998) Diabetes and long-term risk of mortality from coronary and other causes in middle-aged Swedish men. A general population study. Diabetes Care 21(4):539–545

Al-Delaimy WK, Rimm EB, Willett WC (2004) Magnesium intake and risk of coronary heart disease among men. J Am Coll Nutr 23:63–70

Aliabadi H (2008) A deleterious interaction between copper deficiency and sugar ingestion may be the missing link in heart disease. Med Hypotheses 70(6):1163–1166

Chausmer A (1998) Zinc, insulin and diabetes. J Am Coll Nutr 17:109–115

Feroci G, Badiello R, Fini A (2005) Interactions between different selenium compounds and zinc, cadmium and mercury. J Trace Elem Med Biol 18:227–234

Fox CS, Sullivan L, D'Agostino RB Sr, Wilson PW (2004) The significant effect of diabetes duration on coronary heart disease mortality: the Framingham Heart Study. Diabetes Care 27 (3):704–708

Geiss LS, Herman WH, Smith PJ, National Diabetes Data Group (1995) Diabetes in America. National Institutes of Health, National Institute of Diabetes and Digestive and Kidney Diseases, Bethesda, MD, pp 233–257

Gonzalez S, Huerta JM, Alvarez-Uria J, Fernandez S, Patterson AM, Lasheras C (2004) Serum selenium is associated with plasma homocysteine concentrations in elderly humans. J Nutr 134 (7):1736–1740

Helmersson J, Arnlov J, Vessby B, Larsson A, Alfthan G, Basu S (2005) Serum selenium predicts levels of F_2-isoprostanes and prostaglandin F2α in a 27 year follow-up study of Swedish men. Free Radic Res 39:763–770

Howard BV, Lee ET, Yeh JL, Go O, Fabsitz RR, Devereux RB, Welty TK (1996) Hypertension in adult American Indians. Hypertension 28:256–264

International Diabetes Federation (2008) Diabetes prevalence [article online]. Available from http://www.idf.org/home/index.cfm?node=264

Jenner A, Ren B, Rajendran R, Ning P, Huat BTK, Watt F, Halliwell B (2007) Zinc supplementation inhibits lipid per oxidation and the development of atherosclerosis in rabbits fed a high cholesterol diet. Free Radic Biol Med 42:559–566

Kh R, Khullar M, Kashyap M, Pandlu P, Uppal R (2000) Effect of oral magnesium supplementation on blood pressure, platelet aggregation and calcium handling in deoxycorticosterone acetate induced hypertension in rats. J Hypertens 18:919–926

Lefant C, Savage PJ (1995) The early natural history of atherosclerosis and hypertension in the young: National Institutes of Health Perspectives. Am J Med Sci 310(Suppl 1):S3–S7

Legato MJ, Gelze RA, Goland R, Ebner SA, Rajan S, Villagra V, Kosowski M (2006) Gender-specific care of the patient with diabetes: review and recommendations. Gend Med 3:131–158

Linder MC, Hazegh-Azam M (1996) Copper biochemistry and molecular biology. Am J Clin Nutr 63:797S–811S

Linnebank M, Lutz H, Jarre E, Vielhaber S, Noelker C, Struys E (2006) Binding of copper is a mechanism of homocysteine toxicity leading to COX deficiency and apoptosis in primary neurons, PC12 and SHSY-5Y cells. Neurobiol Dis 23(3):725–730

Lissa EM, Bahjri SM, Ahmed WH, Al-Ama N, Ferns GA (2006) Trace element status in Saudi patients with established atherosclerosis. J Trace Elem Med Biol 20:105–114

Ma B, Lawson AB, Liese AD (2006) Dairy, magnesium, and calcium intake in relation to insulin sensitivity: approaches to modeling a dose-dependent association. Am J Epidemiol 164:449–458

Mackay J, Mensah G (2004) Atlas of heart disease and stroke. World Health Organization, Geneva

Mazur A, Maier JAM, Rock E (2007) Magnesium and the inflammatory response: potential physiopathological implications. Arch Biochem Biophys 458:48–56

Miller VT (1990) Dyslipoproteinemia in women. Special considerations. Endocrinol Metab Clin North Am 19(2):381–398

Mostert V (2000) Selenoprotein P: properties, functions and regulation. Arch Biochem Biophys 376:433–438

Navarro-Alarcon M, López-Martínez MC (2000) Essentiality of selenium in the human body: relationship with different diseases. Sci Total Environ 249:347–371

Nayak D, Karmen C, Frishman W, Vakili B (2001) Antioxidant vitamins and enzymatic and synthetic oxygen-derived free radical scavengers in the prevention and treatment of cardiovascular disease. Heart Dis 3(1):28–45

Powell S (2000) The antioxidant properties of zinc. J Nutr 130:1447S–1454S

Rajpathak S, Rimm E, Morris JS, Hu F (2005) Toenail selenium and cardiovascular disease in men with diabetes. J Am Coll Nutr 24:250–256

Ravn-Haren G, Bugel S, Krath BN, Hoac T, Stagsted J, Jorgensen K, Bresson JR, Larsen EH, Dragsted LO (2008) A short-term intervention trial with selenate, selenium-enriched yeast and selenium enriched milk: effects on oxidative defense regulation. Br J Nutr 99:883–892

Rayman MP (2000) The importance of selenium to human health. Lancet 356:233–241

Roussel AM, Kerkeni A, Zouari N, Mahjoub S, Matheau JM, Anderson R (2003) Antioxidant effects of zinc supplementation in Tunisians with type 2 diabetes mellitus. J Am Coll Nutr 22:316–321

Shaper G, Kazantzis G, Schoental R (1979) Cardiovascular disease and trace metals. Proc R Soc Lond B 205:135–143

Shekelle RB, Shryock AM, Oblesby P (1981) Diet, serum cholesterol, and death from coronary heart disease: the Western Electric Study. N Engl J Med 304:65–70

Speich M, Pineau A, Ballereau F (2001) Minerals, trace elements and related biological variables in athletes and during physical activity. Clin Chim Acta 312(1–2):1–11

Suarez A, Pulido N, Casla A (1995) Impaired tyrosine-kinase activity of muscle insulin receptors from hypomagnesaemia rats. Diabetologia 38:1262–1270

Teragawa H, Kato M, Yamagat T, Matsuura H, Kajiyama G (2000) The preventive effect of magnesium on coronary spasm in patients with vasospastic angina. Chest 118:1690–1695

Uusitupa MIJ, Niskanen LK, Siitonen O, Voutilainen E, Pyörälä K (1993) Ten-year cardiovascular mortality in relation to risk factors and abnormalities in lipoprotein composition in type 2 (non-insulin-dependent) diabetic and non-diabetic subjects. Diabetologia 36:1175–1184

Vitale J (1992) Magnesium deficiency and cardiovascular disease. Lancet 340:1224

Vlad M, Berdas E, Tomas R, Sava D, Frakas E, Usa G (1993) Effect of copper sulfate in experimental atherosclerosis. Biol Trace Elem Res 38:47–54

Waśkiewicz A, Piotrowski W, Sygnowska E, Broda G, Drygas W, Zdrojewski T, Kozakiewicz K, Tykarski A, Biela U (2008) Quality of nutrition and health knowledge in subjects with diagnosed cardiovascular diseases in the Polish population – National Multi-centre Health Survey (WOBASZ). Kardiol Pol 66:507–513

Wenger NK (2003) Coronary heart disease: the female heart is vulnerable. Prog Cardiovasc Dis 46:199–229

Whelton PK (1994) Epidemiology of hypertension. Lancet 334:101–106

World Health Organization (WHO) (2009) Fact sheet about cardiovascular disease

Yusuf S, Reddy S, Ôunpuu S, Anand S (2001) Global burden of cardiovascular diseases, part I: general considerations, the epidemiological transition, risk factors, and impact of urbanization. Circulation 104:2746–2753

Chapter 2
Introduction: Lipid Profile

Lipids and lipoproteins are risk factors for CHD. It has been demonstrated that high levels of serum total cholesterol (TC), triglycerides (TG), LDL cholesterol, very-low-density lipoprotein (VLDL), low concentration of HDL cholesterol, and increased body mass index (BMI) are significantly associated with CHD (George and Ludvik 2000). Dyslipidemia is one of the top five major risk factors leading to cardiovascular disorders. It is characterized by elevated LDL cholesterol and TG and decreased HDL cholesterol. Although there are differences in defining dyslipidemia, however, European guideline on CVD prevention in clinical practice recommends TC below 190 mg/dL (5.0 mmol/L) and an LDL cholesterol below 115 mg/dL (3.0 mmol/L) for the general population. The goals are even lower: i.e., <175 mg/dL (4.5 mmol/L) for TC and <100 mg/dL (2.6 mmol/L) for LDL cholesterol in the case of multiple disorders like CHD, other diseases of CVD, or DM (De Backer et al. 2003). Lipid abnormalities significantly contribute to the increased risk of cardiovascular disease and other morbidity in diabetics. VLDL and chylomicrons (CM) are major sources of fatty acid supply to the heart, but little is known about their metabolism in diabetic myocardium. Males and females appear to be equally susceptible to the effects of risk factors such as hypertension, increased plasma LDL cholesterol, and low levels of plasma HDL cholesterol. Estrogens have a favorable effect on lipid profile. It has been observed that they lower LDL cholesterol and elevate HDL cholesterol. Estrogens are thought to increase HDL cholesterol by reducing hepatic triglycerides' lipase activity that catabolizes HDL cholesterol. Global studies of either gender have demonstrated that the risk of atherosclerosis is inversely related to blood levels of HDL cholesterol: the higher the level of HDL cholesterol, the lower will be the risk. It is indicated that for every 1 mg/dL rise in HDL cholesterol, the risk for developing cardiovascular disease decreases by 2–3 % (Toth 2005). HDL cholesterol helps to extract excess cholesterol deposited in blood vessel walls and deliver it back to the liver for elimination through the gastrointestinal tract. HDL cholesterol helps to keep blood vessels dilated, thereby promoting better blood flow. It also reduces blood vessel injury through its antioxidant and anti-inflammatory functions, among other effects. HDL cholesterol carries "old" cholesterol that has been discarded by

N.B. Rizvi and S.A. Nagra, *Minerals and Lipids Profiles in Cardiovascular Disorders in South Asia*, DOI 10.1007/978-3-642-34249-3_2,

cells back to the liver for recycling or excretion. The main function of LDL cholesterol is to transport cholesterol from the liver to the tissues that incorporate it into the cell membranes. The oxidation of LDL cholesterol is believed to have a central role in atherogenesis. Oxidized LDL cholesterol may be involved in atherogenesis by inducing smooth muscle cell proliferation. Acute MI is the most important consequence of coronary artery disease. Some studies have defined TG also as an independent risk factor for MI (Haffner et al. 1998). High TG value could result from the elevation of several lipoproteins such as chylomicrons, different subclasses of VLDL, or intermediate-density lipoproteins (IDL cholesterol). The risk of MI in patients with DM without a history of myocardial infarction is as high as that in patients without MI who have had a myocardial infarction. Mortality after first MI is higher in both males and females with DM. Lipid abnormalities significantly contribute to the increased risk of CVD in diabetes mellitus. Diabetes affects virtually all lipids and lipoproteins. Persons with DM typically have increased plasma concentrations of TG, low plasma concentrations of HDL cholesterol, and slightly raised plasma concentrations of LDL cholesterol. DM is also considered as an independent risk factor for cardiovascular disease (up to fivefold), and as many as 80 % of patients with type II diabetes die from cardiovascular complications (Johnson et al. 2004). Persons with high blood cholesterol levels have a higher prevalence of hypertension, and those with high blood pressure have a higher prevalence of hypercholesterolemia (O'Brien et al. 2003). Abnormalities in plasma lipoprotein metabolism play a central role in the pathogenesis of atherosclerosis, and arterial hypertension with elevated systolic or diastolic blood pressure is positively and independently associated with CHD. The risk of developing CVD associated with the presence of both hypertension and Dyslipidemia has been shown to be greater as compared to hypertension or Dyslipidemia alone (Johnson et al. 2004). Moreover, patients with these two conditions found to have three to four times higher prevalence of MI (Wald and Law 2003). Dyslipidemia causes endothelial damage and consequent loss of physiological vasomotor activity, which may be manifested as increased blood pressure. Asians experience the largest proportion of the worldwide burden of CVD. Further, Asians include several distinct ethnic subpopulations (South Asians, Chinese, etc.), who may differ in their lipid profiles. These differences may be the result of both genetic and environmental factors such as high-cholesterol carbohydrate diets, reduced physical activity, and obesity (Radhika et al. 2009). In a study it has been estimated that LDL-cholesterol, HDL-cholesterol, and TG levels are lower among Asians compared with non-Asians. Also the associations of elevated LDL cholesterol and lower levels of HDL cholesterol for the risk of AMI were found to be broadly similar among Asians and non-Asians.

References

De Backer G, Ambrosioni E, Borch-Johnsen K, Brotons C, Cifkova R, Dallongeville J, Ebrahim S, Faergeman O, Graham I, Mancia G (2003) European guidelines on cardiovascular disease prevention in clinical practice: third joint task force of European and other societies on cardiovascular disease prevention in clinical practice (constituted by representatives of eight societies and by invited experts). Eur J Cardiovasc Prev Rehabil 10(4):S1–S10

George P, Ludvik B (2000) Lipids and diabetes. J Clin Basic Cardiol 3:159–162

Haffner SM, Lehto S, Ronnemaa T, Pyorala K, Laakso M (1998) Mortality from coronary heart disease in subjects with type 2 diabetes and in nondiabetic subjects with and without prior myocardial infarction. N Engl J Med 339:229–234

Johnson ML, Pietz K, Battleman DS, Beyth RJ (2004) Prevalence of comorbid hypertension and dyslipidemia and associated cardiovascular disease. Am J Manag Care 10:926–932

O'Brien E, Asmar R, Beilin L, Imai Y, Mallion JM, Mancia G (2003) European Society of Hypertension recommendations for conventional, ambulatory and home blood pressure measurement. J Hypertens 21:821–848

Radhika G, Ganesan A, Sathya RM, Sudha V, Mohan V (2009) Dietary carbohydrates, glycemic load and serum high-density lipoprotein cholesterol concentrations among South Indian adults. Eur J Clin Nutr 63(3):413–420

Toth PP (2005) The "good cholesterol" high-density lipoprotein. Circulation 111:e89–e91

Wald NJ, Law MR (2003) A strategy to reduce cardiovascular disease by more than 80%. BMJ 326:1419–1423

Chapter 3
Literature Review: Cardiovascular Disorders and Minerals

Tawfeeq et al. (2008) determined the altered status of some essential trace elements and antioxidant minerals in diabetic patients, which could have deleterious influences on the health of diabetics. The mean fasting plasma glucose (FPG) of type I and type II diabetics and BMI of type II diabetics showed statistically significant difference from control group. Mean serum copper of type I and type II diabetics showed statistically significant difference from control group. Simple linear correlation and regression analysis of the FPG level showed strong positive correlation to the Cu and Cu/Zn values in all study groups and negative correlation to the Zn values in all study groups. Also, HbA1c showed strong positive correlation to the Cu and Cu/Zn values among type I diabetics and control groups and negative correlation to the Zn values among type I diabetics and control groups. BMI and duration of diabetes showed no correlation with the Cu, Zn, and Cu/Zn values among all the study groups. It was concluded that serum copper and Cu/Zn ratio increase significantly more than healthy control subjects, while serum zinc showed nonsignificantly reduced levels than healthy control participants, both in type I and II diabetes. Simple linear correlation and regression model analysis showed strong positive correlation of serum copper and Cu/Zn ratio as well as strong negative correlation of zinc toward the FPG values in all studied groups and furthermore toward HbA1c in type I diabetes and control group.

Kazi et al. (2008) determined that the pathogenesis of some heart diseases has been associated with changes in the balance of certain trace elements. They examined the association of iron, copper, and zinc between biological samples (scalp hair, whole blood, and urine) and mortality from myocardial infarction (MI) patients of (first, second, and third heart attack). Seventy-eight percent of the patients aged >50 years registered after the third MI attack died. In these subjects, the concentrations of Fe and Cu were increased by 0.83 % and 3.12 % in the scalp hair, while in blood samples 9.7 % and 22.5 % were enhanced, respectively, as compared to those who tolerated 3rd MI attack ($P = 0.072$). The concentrations of Zn in whole-blood and scalp hair samples were lower in MI patients as compared to normal subjects. Deficiency of zinc and high concentration of copper and iron may play a role in the development of heart disease.

N.B. Rizvi and S.A. Nagra, *Minerals and Lipids Profiles in Cardiovascular Disorders in South Asia*, DOI 10.1007/978-3-642-34249-3_3,

Kazemi-Bajestani et al. (2007) reported an imbalance between zinc and copper metabolism to predispose to coronary artery disease (CAD) in Western populations. They investigated the association between serum copper and zinc and CAD in Iranian subjects undergoing coronary angiography. Male patients had lower serum copper ($P < 0.05$), lower serum zinc ($P < 0.05$), and higher serum zinc/copper ratio ($P < 0.05$) than females. Serum copper and zinc concentrations were significantly lower in the subjects with angiographically defined CAD than those patients with a normal angiogram, although the zinc/copper ratio was higher in these patients ($P < 0.001$). Serum copper ($r = -0.303$, $P < 0.001$) and zinc ($r = -0.250$, $P < 0.01$) concentrations were both inversely related to age, and copper was positively associated with fasting serum triglycerides ($r = 0.188$, $P < 0.05$). They concluded that serum copper and zinc concentrations were significantly lower in Iranian patients with abnormal versus those with a normal angiogram. However, the zinc/copper ratio was higher in patients with CAD compared to subjects without CAD. Serum zinc and copper concentrations appeared to be influenced by several physiological factors including age and gender.

Kosar et al. (2007) determined the serum levels of trace elements, such as selenium, zinc, and copper, in patients with isolated coronary artery ectasia. It is well established that the deficiency of trace elements may lead to oxidative stress in many tissues. They investigated the relationship between the level of trace elements and the extent of ecstatic involvement in patients of coronary artery ectasia. The trace element levels were measured by atomic absorption photometry methods. They found serum selenium (Se) and zinc (Zn) levels in both sets of patients significantly lower than in the control group. However, the serum copper (Cu) levels were similar in all patients and controls. They suggested that coronary artery ectasia is associated with the deficiency of the trace elements selenium and zinc. Thus, these elements may play an important role in the pathogenesis of coronary artery ectasia, as well as in CAD.

Shukla et al. (2006) observed that although ceruloplasmin (CP), a copper containing metalloenzyme, possesses antioxidant properties (e.g., ferroxidase activity), elevated circulating CP is associated with CVD. This ambivalence is possibly due to the capacity of CP, via its coppers, to promote vasculopathic effects that include lipid oxidation, negation of nitric oxide bioactivity, and endothelial cell apoptosis. There is also evidence that risk factors for CVD (in particular, diabetes mellitus, and hyperhomocysteinemia) may augment the vasculopathic impact of ceruloplasmin. In turn, it appears that ROS disrupt copper binding to CP, thereby impairing its normal protective function while liberating copper which in turn may promote oxidative pathology. They concluded that in the healthy individual, copper is tightly bound to CP and its transfer to cells carefully regulated. Although it follows that copper chelators may be indicated as a possible therapeutic intervention for CVD, copper deficiency is also associated with increased risk of atherosclerotic disease, since copper is also crucial for the normal activity of enzymes and systems that counter oxidative stress (e.g., SOD). However, in shorter term clinical scenarios, copper chelators may be therapeutically effective.

Rayman (2005) reported that selenium is an unusual trace element having its own codon in mRNA that specifies its insertion into selenoproteins as selenocysteine (SeCys). Se may protect against cancer; thus, an adequate intake of Se is desirable. The evidence for Se as a cancer preventive agent includes that from geographic, animal, prospective, and intervention studies. Interventions with Se have shown benefit in reducing the risk of cancer incidence and mortality in all cancers combined and specifically in liver, prostate, colorectal, and lung cancers. The effect seems to be strongest in those individuals with the lowest Se status, as the level of Se required for optimal effect is higher than that previously understood to be required to maximize the activity of selenoenzymes. However, recent evidence showing an association between Se and cancer risk implies that selenoproteins are indeed implicated. The likelihood of simultaneous and consecutive effects at different cancer stages still allows an important role for anticancer Se metabolites such as methylselenol formed from gamma-glutamyl-selenomethyl-SeCys and selenomethyl-SeCys, components identified in certain plants and Se-enriched yeast that have anticancer effects. There is some evidence that Se may affect not only cancer risk but also progression and metastasis. Current primary and secondary prevention trials of Se are under way in the USA, including the Selenium and Vitamin E Cancer Prevention Trial (SELECT) relating to prostate cancer, although a large European trial is still desirable given the likelihood of a stronger effect in populations of lower Se status.

Al-Saleh et al. (2005) assessed the status of essential trace elements such as copper, iron, molybdenum, selenium, and zinc in insulin-dependent diabetic pregnancies at term and compared the data with a control group. Values for copper and molybdenum were significantly higher ($P < 0.05$) in the study group compared to control, while those of zinc, iron, and selenium were not significantly different ($P < 0.05$). Iron and molybdenum values were significantly higher ($P < 0.05$) and that of zinc significantly lower ($P < 0.05$) in umbilical arterial samples of diabetic group compared to controls. In the case of molybdenum and copper, the values were significantly higher ($P < 0.05$) in umbilical venous samples of diabetic group compared to that of control. Significant differences in Cu/Zn ratio of maternal venous and umbilical samples and fetal/maternal ratios of some elements were noted between control and study group as well. They speculated that altered status of some essential trace elements and altered antioxidant mineral ratio observed in insulin-dependent diabetic patients could have deleterious influences on the health of the mother as well as the fetus and newborn.

Cooper et al. (2005) showed that treatment with the Cu (II)-selective chelator, trientine, improves left ventricular hypertrophy in humans with type II diabetes and increases urinary Cu excretion in humans compared with nondiabetic control subjects. They characterized the homeostasis of Cu and eight other nutritionally essential elements in diabetes under fully residential conditions in male subjects with type II diabetes and age-matched control subjects. Before treatment, there were no detectable between-group differences in the balance of any element, although urinary output of several elements was greater in diabetic subjects. Mean extracellular superoxide dismutase (EC-SOD) activity was elevated in diabetic subjects, and its activity correlated strongly with the interaction between [Cu] serum and HbA1c. Trientine caused the Cu balance to become negative in diabetic subjects through elevated urinary Cu losses and suppressed elevated EC-SOD. Basal urinary Cu predicted urinary Cu losses during treatment, which caused extraction of systemic Cu (II). They suggested that cardiovascular complications in diabetes might be better controlled by therapeutic strategies that focus on lowering plasma glucose and loosely bound systemic Cu (II).

Pham et al. (2005) observed that hypomagnesemia has been implicated in adversely affecting diabetic complications. They determined an association between serum magnesium concentration [Mg^{2+}] and the rate of renal function deterioration, as determined by the slope of serum creatinine reciprocals versus time, in patients with diabetes mellitus type II (DM). For each patient, all available data from electronic database for [Mg^{2+}], hemoglobin A(1C) (HbA1C), serum creatinine, lipid profiles, routine urinary analysis, as well as history of hypertension and pharmacy profiles were retrieved. Patients were stratified by gender and divided into four groups based on increasing [Mg^{2+}]. They reported that patients belonging to lower [Mg^{2+}] groups for both genders had significantly worse slopes of serum creatinine reciprocals versus time plot (1/SCr-vs-t) independent of the presence of hypertension and use of ACEI/ARB, diuretics, HMG-coA enzyme inhibitors, or aspirin. In a multivariate regression analysis controlling for age, HbA(1C), and various components of the lipid profile, [Mg^{2+}] remained an independent predictor for the slope of 1/SCr-vs-t. A trend for worse proteinuria based on routine urinary analysis was observed among patients belonging to the lowest [Mg^{2+}] group. They concluded that lower [Mg^{2+}] was associated with a faster renal function deterioration rate in type II DM patients.

Ekin et al. (2003) investigated that altered serum total sialic acid (TSA), lipid-associated sialic acid (LSA), copper (Cu), manganese (Mn), zinc (Zn), chromium (Cr), iron (Fe), and magnesium (Mg) levels had an interactive connection with diabetes and also determined that they were correlated with each other in diabetic patients. It was found that diabetics had higher TSA, LSA, Fe, Mn, Fe/Zn, and Cu/Zn levels and lower Zn and Mg levels than those of controls. Although Cu levels were higher and Cr levels were lower in total and male diabetic patients, they were not different in female diabetic patients than in controls. The Cu/Fe ratio was lower in total and female diabetic patients, but not different in male diabetic patients than controls. The Zn/Cr ratio, on the other hand, was not different in diabetics than in controls. There was only a positive correlation between Fe and Mn levels in male diabetic patients. There was a negative correlation in LSA–Mn, Fe–Cu, Cu–Fe/Zn, and Mn–Cu/Zn levels in total diabetic patients. There was a positive correlation in TSA–Cr, TSA–Mg, LSA–Cu/Fe, and LSA–Zn/Cr levels and a negative correlation in TSA–Cu/Zn, LSA–Mn, Fe–Cu, Mn–Cu, Cu–Fe/Zn, Fe–cholesterol, and Cr–cholesterol in female diabetic patients. They concluded that TSA, LSA, and selected minerals have interactive connections with diabetes mellitus (DM). There were also many sex-related positive or negative correlations between the altered parameters in diabetic patients. These parameters might be used as diagnostic index in patients with DM.

Huang et al. (2002) observed that in Se-deficient rats compared with Se-adequate group, the blood Se content, blood and vascular wall glutathione peroxidase (GPx) activity, serum high-density lipoprotein-cholesterol (HDL-cholesterol) level, and plasma prostacyclin (PGI(2)) concentration were decreased significantly, and the blood lipid peroxide (LPO) concentration, serum low-density lipoprotein-cholesterol (LDL-cholesterol) level, total cholesterol (TC) level, and plasma thromboxane A(2) (TXA(2)) content were increased significantly. In severely injured areas, endothelial integrity was completely destroyed, and smooth muscle cells were proliferating and migrated to the endomembrane. They concluded that Se or selenoproteins in the vascular wall play an important role in cytoprotection against cholesterol oxide-induced vascular damage in rats.

Institute of Medicine (2002) provided quantitative reference intakes for vitamin A, vitamin K, boron, chromium, copper, iodine, iron, manganese, molybdenum, nickel, vanadium, and zinc. No recommendations were provided for arsenic and silicon. The development of Dietary Reference Intakes expands and replaces the series of Recommended Dietary Allowances (RDAs) in the United States and Recommended Nutrient Intakes in Canada. The growing recognition of the many uses to which RDAs and Recommended Nutrient Intakes have been applied and the growing awareness that many of these uses require the application of statistically valid methods that depend on reference values other than RDAs and Recommended Nutrient Intakes. This report included a review of the roles that micronutrients are known to play in traditional deficiency diseases and evaluates possible roles in chronic diseases.

Stühlinger (2002) reported that in cardiovascular medicine, magnesium is of major importance in the treatment of arrhythmias and coronary artery disease. Magnesium raises the ventricular fibrillation threshold and prolongs the sinus node recovery time and atrioventricular conduction time. Magnesium has been used successfully in ventricular ectopies after overdose of neuroleptics or tricyclic antidepressants. Potential benefits can be expected in monomorphic ventricular tachycardias and in ventricular arrhythmias that did not respond to class III antiarrhythmic drugs. Recent studies have shown positive effects of magnesium in preoperative patients, where the incidence of atrial and ventricular arrhythmias could be reduced. Oral magnesium has been used for years in patients with premature ventricular beats. Several studies have shown that combined oral therapy with magnesium and potassium can effectively reduce the incidence of premature ventricular beats. Patients with coronary heart disease frequently suffer from magnesium deficiency. Oral combination therapy with magnesium and potassium improves the endothelial function in these patients and reduces platelet-dependent thrombosis. These encouraging results from basic science studies have now been confirmed in a large clinical trial showing that oral magnesium therapy improves exercise duration and quality of life in patients with coronary artery disease.

Elwood and Pickering (2002) reported magnesium (Mg) as an essential element with numerous biological functions. Its relevance to disease is, however, uncertain. Both dietary Mg intake and serum Mg appeared to be negatively related to the incidence of ischemic heart disease events and stroke, but the evidence is both sparse and inconsistent. Perhaps the most persuasive data come from the postmortem examinations of heart muscle: low tissue Mg levels being consistently found in myocardium taken from subjects whose death had been attributed to vascular disease. The clinical aspect which has received the greatest attention in recent years is the value of an infusion of Mg given early to patients with acute myocardial infarction. The evidence on this is inconsistent, and a further trial is being set up in an attempt to resolve some of the uncertainties. The most pressing needs in the evaluation of Mg and health are, however, for more prospective evidence on dietary Mg intake and the risk of myocardial infarction and for randomized controlled trials of the effect of an increase in Mg intake on vascular disease mortality.

Rayman (2000) observed the fundamental importance of selenium to human health. As a constituent of selenoproteins, selenium has structural and enzymatic roles as an antioxidant and catalyst for the production of active thyroid hormone. Se is needed for the proper functioning of the immune system and appears to be a key nutrient in counteracting the development of virulence and inhibiting HIV progression to AIDS. Deficiency has been linked to adverse mood states. Findings have been equivocal in linking Se to CVD risk, although other conditions involving oxidative stress and inflammation have shown benefits of a higher Se status. An elevated Se intake may be associated with reduced cancer risk. In the context of these health effects, low or diminishing selenium status in some parts of the world, notably in some European countries, is giving cause for concern.

Saris et al. (2000) observed an increased interest in the role of magnesium ions in clinical medicine, nutrition, and physiology. Magnesium affects many cellular functions, including transport of potassium and calcium ions, and modulates signal transduction, energy metabolism, and cell proliferation. The mechanism of cellular uptake and efflux of magnesium, its intracellular transport, intestinal absorption, and renal excretion, and the effect of hormones on these were reviewed. Magnesium deficiency is not uncommon among the general population: its intake has decreased over the years especially in the Western world. The magnesium supplementation or intravenous infusion may be beneficial in various diseased states. Of special interest is the magnesium status in alcoholism, hypertension, atherosclerosis, cardiac diseases, diabetes, and asthma. The improved procedures for the assay of different magnesium states are useful in understanding the role of magnesium in health and disease.

Kruse-Jarres and Rükgauer (2000) observed that significantly more information about trace element status can be obtained by investigating concentrations in blood cells instead of only evaluating the concentrations in plasma. Essential trace elements such as zinc, copper, chromium, and selenium take part in a variety of enzymatic processes on a molecular–cellular level. Especially in metabolic diseases like diabetes mellitus, conclusions drawn from trace element concentrations in blood cells usually offer more valuable clinical information about the metabolic state than trace element concentrations in plasma or whole blood. In the present investigation, copper and zinc concentrations were increased in all blood fractions of diabetic patients (IDDM). In insulin-dependent diabetic children, significantly higher values of zinc in erythrocytes were also found, and they were higher in patients with poor metabolic control (HbA1c > 9 %). When different blood fractions in diabetic patients (NIDDM) were compared with a control group, chromium was significantly increased in plasma and polymorphonuclear cells. Patients with IDDM had pronounced decreased selenium concentrations in erythrocytes as compared to controls.

Meerarani et al. (2000) reported that zinc requirements of the vascular endothelium may be increased in inflammatory conditions, that is, atherosclerosis, in which apoptotic cell death is prevalent. They hypothesized that zinc deficiency may potentiate disruption of endothelial cell integrity mediated by fatty acids and inflammatory cytokines by enhancing pathways that lead to apoptosis and upregulation of caspase genes. Supplementation of low-serum- or chelator-treated endothelial cells with physiological amounts of zinc caused a marked attenuation of apoptosis induced by linoleic acid and TNF-α. Morphologic changes of cells observed during zinc deficiency were prevented by zinc supplementation. Media supplementation with other divalent cations (e.g., calcium and magnesium) did not mimic the protective role of zinc against apoptosis. They concluded that zinc is vital to vascular endothelial cell integrity, possibly by regulating signaling events to inhibit apoptotic cell death.

Jack T. Saari (2000) reported that dietary copper deficiency causes a variety of cardiovascular deficits. Systemic effects include high blood pressure, enhancement of inflammation, anemia, reduced blood clotting, and possibly arteriosclerosis. Effects on specific organs or tissues include weakened structural integrity of the heart and blood vessels, impairment of energy use by the heart, reduced ability of the heart to contract, altered ability of blood vessels to control their diameter and grow, and altered structure and function of circulating blood cells. In some instances, the cause of a defect can be directly attributed to reduced activity of a specific copper-dependent enzyme. However, three nonspecific mechanisms of damage have been implicated in cardiovascular defects of copper deficiency. They are peroxidation, the interaction of oxygen-derived free radicals with lipids and proteins; glycation, the nonenzymatic glycosylation of proteins; and nitration, the interaction of nitric oxide and its metabolites with peptides and proteins. Though independently these mechanisms present great potential for damage, the possibility that they may interact presents an added reason for concern. Furthermore, the fact that at least two of these mechanisms are associated with diabetes and aging suggests that copper deficiency may exacerbate deficits associated with these two conditions.

Ma and Betts (2000) determined that zinc and copper are two trace minerals essential for important biochemical functions and necessary for maintaining health throughout life. Several national food surveys revealed marginally to moderately low contents of both nutrients in the typical American diet. The daily zinc intake was 12 ± 6.4 mg for men and 8.0 ± 4.0 mg for women ($P < 0.05$); the daily copper intake was 1.3 ± 0.7 mg for men and 1.0 ± 0.5 mg for women ($P < 0.05$). Foods such as beef, ground beef, legumes, poultry, ready-to-eat and hot cereals, and pork constituted the major sources of zinc. Copper consumption was contributed mainly by legumes, potato and potato products, nuts and seeds, and beef. The less-than-recommended intakes of zinc and copper by the elderly were likely associated with age, low income, and less education. The intakes of zinc and copper could be improved by more frequent consumption of food sources rich in these minerals. An inherent limitation of this study was the use of the 24-h dietary recall method, which may underestimate usual dietary intakes. Nonetheless, this study affirms the need for assessment of zinc and copper nurture in the elderly.

Ford (1999) reported that magnesium may play an important role in ischemic heart disease. A few prospective epidemiological studies have related serum magnesium concentrations to mortality from ischemic heart disease (IHD) or all-causes. Data from the National Health and Nutrition Examination Survey Epidemiologic Follow-up Study were used to examine the association between serum magnesium concentration, measured between 1971 and 1975, and mortality from IHD or all-causes in a national sample of 25- to 74-year-old participants followed for about 19 years. No significant interactions between serum magnesium concentration and age, sex, race, and education were observed. They concluded that serum magnesium concentrations were inversely associated with mortality from IHD and all-cholesterol use mortality.

Hennig et al. (1999) observed that little is known about the requirements and function of zinc in maintaining endothelial cell integrity, especially during stressful conditions, such as the inflammatory response in cardiovascular disease. Zinc requirements of the vascular endothelium are increased during inflammatory conditions such as atherosclerosis, where apoptotic cell death is also prevalent. Apoptosis is a morphologically distinct mechanism of programmed cell death which involves the activation of a cell-intrinsic suicide program, and there is evidence that factors such as inflammatory cytokines (e.g., tumor necrosis factor [TNF]) and pure or oxidized lipids are necessary to induce the cell death pathway. Because of its constant exposure to blood components, including prooxidants, diet-derived fats, and their derivatives, the endothelium is very susceptible to oxidative stress and to apoptotic injury mediated by blood lipid components, prooxidants, and cytokines. Thus, it is likely that the cellular lipid environment, primarily polyunsaturated fatty acids, can potentiate the overall endothelial cell injury by increasing cellular oxidative stress and cytokine release in proximity to the endothelium, which then could further induce apoptosis and disrupt endothelial barrier function. They suggested that zinc deficiency exacerbates the detrimental effects of specific fatty acids (e.g., linoleic acid) and inflammatory cytokines, such as TNF, on vascular endothelial functions. They proposed that a major mechanism of zinc protection against disruption of endothelial cell integrity during inflammatory conditions was by the ability of zinc to inhibit the pathways of signal transduction leading to apoptosis and especially mechanisms that lead to upregulation of caspase genes.

Suter (1999) reported that increased potassium intake may play a role in the prevention and treatment of hypertension. They also discussed possible mechanisms by which potassium may act as an antihypertensive and recommendations regarding increasing potassium intake. Stroke mortality represents the third leading cause of death worldwide, after coronary artery disease and cancer. High blood pressure is a major risk factor for stroke. Potassium, magnesium, and fiber have been identified as significant modulators of stroke risk in men. The protective effects were particularly pronounced in hypertensive subjects. The observed protection may be due to direct and indirect effects of these nutrients on blood pressure and regulatory functions, such as endothelial function. A high intake of these nutrients, singularly or in combination, is associated with a more healthful overall lifestyle. The best strategy to achieve a high intake of these nutrients is a diet rich in fruits and vegetables.

Zargar et al. (1998) reported a relationship between trace elements and diabetes mellitus. They evaluated the role of such a relationship in patients with non-insulin-dependent diabetes mellitus with a mean duration of diabetes of 3.9 ± 3.6 years. Diabetic subjects were also subdivided into controlled and uncontrolled groups; control was based on fasting blood glucose and serum fructosamine levels. Plasma zinc and magnesium levels were comparable between diabetic and nondiabetic subjects, while copper levels were significantly elevated ($P < 0.01$) in diabetic patients. Age, sex, duration, and control of diabetes did not influence copper, zinc, or magnesium concentrations. They concluded that zinc and magnesium levels were not altered in diabetes mellitus, but the increased copper levels found in diabetics in their study may merit further investigation of the relationship between copper and non-insulin-dependent diabetes mellitus.

Singh et al. (1998) determined the association between current zinc intake and prevalence of coronary artery disease (CAD) and diabetes as well as factors associated with insulin resistance. The prevalence of CAD, diabetes, and glucose intolerance was significantly higher among subjects consuming lower intakes of dietary zinc. There was a higher prevalence of hypertension, hypertriglyceridemia, and low high-density lipoprotein-cholesterol levels which showed significant upward trend with lower zinc intakes. Serum lipoprotein(a) and plasma insulin levels also were associated with low zinc intake. Serum zinc (odds ratio: men 0.77, women 0.57), serum triglycerides (men 0.86, women 0.81), blood pressure (0.83 men, women 0.76), diabetes mellitus (men 0.90, women 0.85), central obesity (men 0.88, women 0.87), glucose intolerance (men 0.66, women 0.57), and low high-density lipoprotein cholesterol (men 0.72, women 0.70) were significant risk factors for CAD (explained by tertiles of zinc status) in urban subjects. These associations were not observed in rural subjects. They concluded that lower consumption of dietary zinc and low serum zinc levels were associated with an increased prevalence of CAD and diabetes and several of their associated risk factors including hypertension, hypertriglyceridemia, and other factors suggestive of mild insulin resistance in urban subjects.

Singh et al. (1998) determined the association between current zinc intake and prevalence of coronary artery disease (CAD) and diabetes as well as factors associated with insulin resistance. They observed that the prevalence of CAD, diabetes, and glucose intolerance was significantly higher among subjects consuming lower intakes of dietary zinc. There was a higher prevalence of hypertension, hypertriglyceridemia, and low HDL-cholesterol levels which showed significant upward trend with lower zinc intakes. Serum lipoprotein(a) and plasma insulin levels also were associated with low zinc intake. Multivariate logistic regression analysis after adjustment for age showed that zinc intake and CAD were inversely associated. Serum zinc (odds ratio: men 0.77, women 0.57), serum triglycerides (men 0.86, women 0.81), blood pressure (0.83 men, women 0.76), diabetes mellitus (men 0.90, women 0.85), central obesity (men 0.88, women 0.87), glucose intolerance (men 0.66, women 0.57), and low high-density lipoprotein cholesterol (men 0.72, women 0.70) were significant risk factors for CAD in urban subjects. These associations were not observed in rural subjects. They concluded that lower consumption of dietary zinc and low serum zinc levels were associated with an increased prevalence of CAD and diabetes and several of their associated risk factors including hypertension, hypertriglyceridemia, and other factors suggestive of mild insulin resistance in urban subjects.

Choi (1998) determined zinc as an essential catalytic or structural element of many proteins and a signaling messenger that is released by neural activity at many central excitatory synapses. Growing evidence suggested that zinc may also be a key mediator and modulator of the neuronal death associated with transient global ischemia and sustained seizures, as well as perhaps other neurological disease states. Manipulations aimed at reducing extracellular zinc accumulation, or cellular vulnerability to toxic zinc exposure, may provide a novel therapeutic approach toward ameliorating pathological neuronal death in these settings.

Metwalli et al. (1998) studied the changes of serum lipid and metal levels in acute myocardial infarction (AMI) patients in comparison to normal subjects. The same parameters were also determined in presence of other diseases in conjunction to AMI such as diabetes and/or hypertension. They showed that all patients' magnesium, copper, triglyceride, and low-density lipoprotein-cholesterol (LDL-cholesterol) levels were significantly higher than the corresponding values of controls. Serum high-density lipoprotein-cholesterol (HDL-cholesterol) levels were lower than values of controls, but the decrease was significant in patients with AMI and AMI with diabetes. Also, serum iron significantly decreased in patients with AMI, diabetes, and hypertension. In all patients, serum zinc and total cholesterol (T cholesterol) showed an appreciable increase compared to the control. Their findings emphasized the importance of lipid and metal measurements in AMI patients.

Liao et al. (1998) hypothesized that hypomagnesemia plays a role in coronary heart disease (CHD), but few prospective epidemiological studies have been conducted. They examined the relation of serum and dietary magnesium with CHD incidence in a sample of middle-aged adults ($n = 13{,}922$ free of baseline CHD) from four US communities. Over 4–7 years of follow-up, a few men and women had CHD develop. After adjustment for sociodemographic characteristics, waist/hip ratio, smoking, alcohol consumption, sports participation, use of diuretics, fibrinogen, total cholesterol and high-density lipoprotein-cholesterol levels, triglyceride levels, and hormone replacement therapy, the relative risk of CHD across quartiles of serum magnesium was 1.00, 0.92, 0.48, and 0.44 (P for trend $= 0.009$) among women and 1.00, 1.32, 0.95, and 0.73 (P for trend $= 0.07$) among men. The adjusted relative risk of CHD for the highest versus the lowest quartile of dietary magnesium was 0.69 in men (95 % confidence interval

0.45–1.05) and 1.32 in women (0.68–2.55). These findings suggested that low magnesium concentration may contribute to the pathogenesis of coronary atherosclerosis or acute thrombosis.

Rude (1998) reported that magnesium is the most prevalent intracellular divalent cation and the second most prevalent cation in the body. The normal adult body content is ~25 g, and its distribution is approximately equally divided between the skeleton and soft tissues. The physiological role of Mg is principally related to enzyme activity; over 300 enzyme systems are dependent on the presence of this cation. Efficient mechanisms exist in both the gastrointestinal tract and the kidney that closely regulate Mg homeostasis. In the intestine, an active Mg transport system accounts for greater fractional Mg absorption at low dietary intake. Caution should be used with Mg therapy in patients with any degree of renal failure because hypomagnesemia may develop. If a decrease in the glomerular filtration rate exists, the dose of Mg should be halved, and the serum Mg concentration must be monitored daily. If hypomagnesemia ensues, therapy must be stopped. If the patient has normal renal function, the excess Mg will be excreted into the urine. Patients with severe Mg intoxication may be treated with intravenous calcium. The usual dose is an infusion of 100–200 mg of elemental calcium over 5–10 min. Dialysis may be required if the patient is in renal failure.

Martin-Lagos et al. (1997) evaluated serum copper and zinc levels in patients with cancer (respiratory, digestive, hematological, and gynecological) and patients with cardiopathy (acute myocardial infarction and ischemic cardiomyopathy). A control group of healthy subjects was also selected. The mean serum zinc levels in patients with gynecological cancer and ischemic cardiomyopathy were significantly lower than the control group ($P < 0.05$). However, the mean serum copper level was not statistically different among patients with cancer ($P > 0.05$) and cardiomyopathy ($P > 0.05$) than the control group. Male patients did not have statistically different values for serum Cu ($P > 0.05$) and Zn ($P > 0.05$) than those found in female patients. Patients' age did not have any statistical influence ($P > 0.05$) on serum Cu and Zn levels.

Vilanova et al. (1997) reported that metallothionein (MT) is a protein characterized by a high metal-binding capacity. Variations in metal concentrations may occur in several pathologies such as acute myocardial infarction (AMI). They investigated Zn, Cu, and MT levels in patients affected by AMI and also studied their possible role in the patient's short-term survival. Zn and Cu serum levels were determined by atomic absorption spectrophotometer, and MT concentrations were measured by RIA. A significant increase in serum Cu and MT levels and a decrease in Zn levels in the AMI group were observed. A significant decrease in serum Zn and Cu levels in the subset of deceased patients was obtained. They concluded that MT behaves as an acute phase protein in AMI, though it is not a proper prognosis marker of this disease in the short term, and it does not seem to be related to Zn and Cu serum levels.

Neve (1996) examined selenium as a powerful antioxidant, regulating the activity of the glutathione peroxidase enzymes, which catalyze the detoxification of hydrogen peroxide and organic hydroperoxides. Selenium deficiency has been implicated in congestive cardiomyopathy in subjects on artificial nutrition. However, the evidence from case–control and prospective studies for an association between low selenium status and cardiovascular diseases remains controversial. Mechanisms whereby selenium protects against such diseases include increased resistance of low-density lipoproteins against oxidative modification, modulation of prostaglandin synthesis and platelet aggregation, and protection against toxic heavy metals. The therapeutic benefit of selenium administration in the prevention and treatment of cardiovascular diseases still remains insufficiently documented.

Geleinjoise et al. (1996) examined the relation between dietary electrolyte intake and blood pressure in older people. An increase in potassium intake of 1 g/day was associated with a 0.9 mmHg lower systolic and a 0.8 mmHg lower diastolic blood pressure. An increase in magnesium intake of 100 mg was associated with a 1.2 mmHg lower systolic and a 1.1 mmHg lower diastolic blood pressure. Calcium intake was not independently related to blood pressure, except for a subgroup of hypertensive subjects in which a significant inverse association with diastolic blood pressure was observed. Their findings supported the view that an increase in the intake of foods rich in potassium and magnesium could lower blood pressure at older age.

Ascherio et al. (1996) examined prospectively the relation of nutritional factors with hypertension and blood pressure levels among predominantly white US female nurses, aged 38–63 years, and without diagnosed hypertension, cancer, or cardiovascular disease. Some of the women reported a diagnosis of hypertension. Age, relative weight, and alcohol consumption were the strongest predictors for the development of hypertension. Dietary calcium, magnesium, potassium, and fiber were not significantly associated with risk of hypertension, after adjusting for age, body mass index, alcohol, and energy intake. Among women who did not report hypertension during the follow-up period, calcium, magnesium, potassium, and fiber were each significantly inversely associated with self-reported systolic and diastolic pressures, after adjusting for age, body mass index, alcohol consumption, and energy intake. In analyses of food groups, intakes of fruit and vegetables were inversely associated with systolic and diastolic pressures, and intakes of cereals and meat were directly associated with systolic pressure. These results supported the hypotheses that age, body weight, and alcohol consumption are strong determinants of risk of hypertension in middle-aged women. They are compatible with the possibilities that magnesium and fiber as well as a diet richer in fruits and vegetables may reduce blood pressure levels.

Hennig et al. (1996) reported zinc as an essential component of biomembranes and is necessary for maintenance of membrane structure and function. There is evidence that zinc can provide anti-atherogenic properties by preventing metabolic physiological derangements of the vascular endothelium. Because of its antioxidant and membrane-stabilizing properties, zinc appears to be crucial for the protection against cell-destabilizing agents such as polyunsaturated lipids and inflammatory cytokines. Zinc also may be anti-atherogenic by interfering with signaling pathways involved in apoptosis. Most importantly, they evidence that zinc can protect against inflammatory cytokine-mediated activation of oxidative stress-responsive transcription factors, such as nuclear factor-κB and AP-1. It is very likely that certain lipids and zinc deficiency may potentiate the cytokine-mediated inflammatory response and endothelial cell dysfunction in atherosclerosis. Thus, the anti-atherogenic role of zinc appears to be in its ability to inhibit oxidative stress-responsive factors involved in disruption of endothelial integrity and atherosclerosis. They discussed anti-atherogenic properties of zinc with a focus on endothelial cell metabolism.

Linder and Hazegh-Azam (1996) reviewed the basic and most recent understanding of copper biochemistry and molecular biology for mammals (including humans). Information was provided on the nutritional biochemistry of copper, including food sources, intestinal absorption, transport, tissue distribution, and excretion, along with descriptions of copper-binding proteins and other factors involved and their roles in these processes. Alterations in copper metabolism associated with genetic and nongenetic diseases were summarized, including potential connections to inflammation, cancer, atherosclerosis, and anemia, and the effects of genetic copper deficiency (Menkes syndrome) and copper overload (Wilson disease). Understanding these diseases suggested new ways of viewing the normal functions of copper and provided new insights into the details of copper transport and distribution in mammals.

Holecek et al. (1995) observed that many free radicals are formed in diabetes mellitus due to the oxidative stress in hyperglycemia. Apolipoprotein A increased significantly ($P < 0.05$). Glycosylated hemoglobin in the diabetic group decreased significantly from 9.4 ± 1.6 to 7.2 ± 1.5 μmol fructose/g of hemoglobin ($P < 0.01$). In the diabetic group, levels of inorganic phosphorus increased from 0.99 ± 0.08 to 1.15 ± 0.13 mmol/L ($P < 0.01$), zinc from 10.4 ± 1.3 to 14.3 ± 1.7 μmol/L ($P < 0.01$), copper from 20.3 ± 2.3 to 25.9 ± 6.3 μmol/L ($P < 0.05$), and selenium in blood from 0.96 ± 0.21 to 1.65 ± 0.38 μmol/L ($P < 0.001$). Selenium in blood of the control group increased from 0.88 ± 0.26 to 1.66 ± 0.34 μmol/L ($P < 0.001$). The activity of superoxide dismutase (SOD) in diabetic group increased from 598 ± 105 to 696 ± 103 U/g of hemoglobin ($P < 0.01$), the activity of glutathione peroxidase (GSHPx) did not change, malondialdehyde (MDA) decreased from 7.1 ± 1.1 to 5.8 ± 1.1 μmol/L ($P < 0.01$), and uric acid decreased from 261 ± 83 to 236 ± 96 μmol/L ($P < 0.01$). They concluded that multivitamin mixture with trace elements significantly protects diabetic patients and the control group against injurious actions of free radicals. That is confirmed by the decrease of plasmatic malondialdehyde and uric acid and by the increase of superoxide dismutase in erythrocytes. The decrease of glycosylated hemoglobin reduced the probability of diabetic complications; the increase of plasmatic Zn and Se in the diabetic group increased the plasmatic antioxidant ability.

Wang et al. (1995) investigated the effect of increased selenium uptake on serum selenium in diabetic children during the first 9 years of the Finnish nationwide selenium fertilization program. Serum selenium concentrations were followed in diabetic children (mean age 8.1 years) and controls from 1984 to 1992. The effect of the increased uptake was seen in both diabetic and in control persons. Diabetic patients had significantly higher serum selenium levels than their siblings or the other healthy controls. Toward the end of year 1987, this difference had disappeared. After that, serum selenium levels continued to increase until the year 1990. In 1990, the mean selenium serum level of diabetic patients was 1.36 μmol/L and that of controls 1.33 μmol/L. The duration of diabetes did not have any effect on selenium serum levels. Slightly higher serum selenium in new diabetic patients before the start of therapy was explained by the dehydration state. The patients who were younger than 3 years had slightly lower selenium serum levels when compared with older age groups. This difference was observed, however, only during the first 3 years of the study. After that, when the selenium intake increased in general, no age-dependent differences were found anymore. There were no significant differences in serum selenium levels between males and females in either diabetic patients or in controls.

Maj-Zurawska (1994) investigated the influence of anion ligands on Mg–ISE responses and measured ionized magnesium colorimetrically. Ionized magnesium (iMg) concentration changed with pH. The dependence of iMg and iCa on heparin (used as anticoagulant in plasma samples) was made. The range of iMg and total Mg in serum and plasma of healthy and ill adults (short bowel syndrome, myocardial infarction) was investigated and compared to each other. In both illnesses, the concentration of ionized and/or total magnesium was significantly different from the range for healthy people. The range of iMg and total Mg as well as the percentage of iMg against total Mg seemed to be specific for given illness. They also investigated the range of iMg and total Mg in blood serum of healthy children.

Milne and Johnson (1993) measured major indices related to copper nutritional status in men and women between ages 20 and 83 years who were in apparent good health. Plasma copper concentrations and both immunoreactive and enzymatically measured ceruloplasmin were significantly higher in women than in men and were higher in women taking oral contraceptives. Plasma copper, immunoreactive ceruloplasmin, and cytochrome–cholesterol oxidase in platelets and mononucleated leukocytes tended to increase with age. The ratio of enzymatic to immunoreactive ceruloplasmin, erythrocyte superoxide dismutase, and 67Cu uptake by erythrocytes was not significantly affected by either age or gender. They concluded that factors other than copper nurture—such as age, gender, and hormone use—need to be considered when using many of these indicators to evaluate copper nutritional status.

Resnick et al. (1993) investigated the alterations of magnesium metabolism in type II (non-insulin-dependent) diabetes mellitus. They utilized a new magnesium-specific selective ion electrode apparatus to measure serum-ionized magnesium (Mg-io) in fasting subjects with and without type II diabetes and compared these values to levels of serum total magnesium and of intracellular free magnesium (Mgi). Both Mg-io (0.630 ± 0.008 vs. 0.552 ± 0.008 mmol/L, $P < 0.001$) and Mgi (223.3 ± 8.3 vs. 184 ± 13.7 mmol/L, $P < 0.001$), but not serum total magnesium, were significantly reduced in type II diabetes compared with nondiabetic control subjects. Furthermore, a close relationship was observed between serum Mg-io and Mgi ($r = 0.728$, $P < 0.001$). They suggested that magnesium deficiency, both extracellular and intracellular, was a characteristic of chronic stable mild type II diabetes and as such may predispose to the excess cardiovascular morbidity of the diabetic state. Furthermore, by more adequately reflecting cellular magnesium metabolism than total serum magnesium levels, Mg-io measurements may provide a more readily available tool than has heretofore been available to analyze magnesium metabolism in a variety of diseases.

Akyüz et al. (1993) studied serum magnesium, zinc, copper, and ascorbic acid in subjects with atherosclerotic heart disease (AHD), patients with hypertension, patients who had atherosclerotic heart disease with hypertension, and healthy controls. Mg levels were significantly lower in all the study groups compared with controls ($P < 0.001$); however, Cu levels were significantly higher in all three study groups compared with controls ($P < 0.01$). Zn levels had no statistical significance compared to controls. AHD group and AHD with hypertension group had significantly lower ascorbic acid levels compared with controls ($P < 0.01$). However, the difference found in the hypertensive group was not statistically significant ($P > 0.05$). Besides the functional and biochemical risk factors, we conclude that Mg, Zn, and Cu levels and ascorbic acid deficiency may have an important role in pathogenesis of hypertension and AHD.

Landmark and Urdal (1993) estimated magnesium in patients with acute myocardial infarction (AMI) and divided them into three groups according to the time elapsed from onset of chest pain to when the first sample for determination of magnesium (s-Mg1) and potassium (s-K1) was drawn. Ten patients in group A receiving nonselective beta-blockers had an attenuated drop in s-Mg1, whereas the drop in s-K1 was completely inhibited. The differences between s-Mg2 and s-Mg1 (delta s-Mg) in all groups and between s-K2 and s-K1 (delta s-K) in group A increased with increasing mean peak creatine kinase (CKmax) levels to approximately 1,300–1,800 U/L. In conclusion, these observations suggested that the initial drop in s-Mg and s-K in the early phase of AMI is due to increased stimulation of beta-2 adrenergic receptors; these changes can be prevented partly or completely by the use of nonselective beta-blockers.

McMaster et al. (1992) measured serum copper and zinc concentrations (μmol/L) in non-fasting subjects aged 25–64 years participating in two independent population surveys in Northern Ireland. In 1983–1984, copper in males was 17.2 ± 3.1 (mean ± SD) and zinc was 12.1 ± 1.7 (SD). Copper in females, neither pregnant nor taking estrogens nor progestogens, was 19.0 ± 3.9 and zinc was 11.6 ± 1.4. In 1986–1987, copper in males was 17.9 ± 3.3 and zinc was 13.2 ± 2.1. Copper in females was 20.1 ± 3.9 and zinc was 12.7 ± 2.0. Zinc but not copper concentrations decreased from early morning to late afternoon; both were unaffected by reported postprandial time. There was a positive relationship between copper and age for both sexes, but zinc showed only a slight upward trend with age. A positive relationship between copper and the aggregation of classical risk factors for coronary heart disease was demonstrated.

Schnack et al. (1992) reported frequently low levels of magnesium in diabetes mellitus especially in poorly controlled type I (insulin-dependent) diabetic patients. Furthermore, hypomagnesemia might contribute to insulin resistance in type II (non-insulin-dependent) diabetes. As the influence of improved metabolic control on plasma magnesium levels was unknown in type II diabetic patients, they studied magnesium plasma levels in patients before, 1 and 3 months after the initiation of insulin therapy or intensified treatment with oral hypoglycemic agents. Magnesium plasma levels were measured by a colorimetric method and were significantly reduced in diabetic patients compared to healthy control subjects (0.79 ± 0.01 mmol/L vs. 0.88 ± 0.01 mmol/L; $P < 0.0001$). Metabolic control was significantly improved as documented by reduced HbA_{1C} levels in either insulin-treated patients or the patients on oral hypoglycemic agents ($P < 0.003$). However, plasma magnesium levels remained unchanged during the follow-up in the insulin-treated group (1:0.79 ± 0.02 mmol/L; 2:0.81 ± 0.02 mmol/L; 3:0.79 ± 0.01 mmol/L) as well as in the patients on oral hypoglycemic agents (1:0.79 ± 0.03 mmol/L; 2:0.78 ± 0.02 mmol/L; 3:0.84 ± 0.04 mmol/L). They showed that even marked improvement of glycemic control does not correct hypomagnesemia in type II diabetes. They concluded that hypomagnesemia might be related to the insulin-resistant state and that possible beneficial effect of chronic magnesium administration should be evaluated in these patients.

Ferrara et al. (1992) evaluated the effects of magnesium pidolate (15 mmol/day) on blood pressure at rest and during sympathetic stimulation induced by cold, isometric, and tilt test; peripheral blood flow by strain-gauge plethysmography. In the actively treated group, magnesium urinary excretion increased from 5.3 ± 2 to 7.7 ± 2 mmol/24 h, and serum magnesium changed from 0.9 ± 0.1 to 1.0 ± 0.2 mmol/L. On magnesium, BP changed at rest from 156/97 ± 12/4 to 149/90 ± 8/3 mmHg, during cold pressor test from 169/105 ± 9/6 to 174/105 ± 15/4, during isometric exercise from 170/107 ± 13/9 to 170/105 ± 20/6, and during tilt test from 149/96 ± 11/6 to 153/96 ± 17/7 mmHg. Similar changes were observed in the placebo group. Peripheral resistances were 14.7 ± 4 and 9.8 ± 2 PRU before and after magnesium, respectively. These data indicated that long-term magnesium pidolate supplementation does not affect blood pressure at rest and during sympathetic stimulation, despite a slight, nonsignificant reduction in forearm peripheral resistance.

Nadler et al. (1992) evaluated the intracellular erythrocyte magnesium concentration in non-insulin-dependents (type II) diabetics. Magnesium deficiency may be an important factor leading to CVD. Diabetic subjects showed an increase in platelet reactivity that could enhance the risks of vascular disease. In addition, diabetic patients had been reported to be at risk of developing extracellular magnesium deficiency. However, the intracellular free magnesium concentration and its role in the enhanced platelet reactivity in diabetes are not known. They found intracellular magnesium concentration of diabetic patients significantly reduced compared with nondiabetic control subjects. Serum magnesium levels were also found to be reduced in the diabetic patients compared with the control subjects. Both intravenous and oral magnesium supplementation markedly reduced platelet reactivity. The magnesium-free diet resulted in a significant reduction in magnesium concentration and markedly enhanced the sensitivity of platelet aggregation. They suggested that type II diabetic patients had intracellular magnesium deficiency and that magnesium deficiency may be a key factor in leading to enhanced platelet reactivity in type II diabetes. Therefore, magnesium supplementation may provide a new therapeutic approach to reducing vascular disease in patients with diabetes.

Neve (1991) assessed selenium status in man by determination of the selenium concentration in biological materials and by the measurement of biochemical or clinical selenium-dependent functions. Over and above the discussion of normal levels and of the numerous factors liable to influence the above-mentioned indices, the present review more specifically deals with the sensitivity of various parameters to changing selenium status, both in deficiency states and in overexposure to the element, and the attempt was made to establish the specificity of these indicators in the diagnosis of selenium states. The possibilities for defining "adequate" or "optimum" selenium concentrations were also examined.

Salonen et al. (1991) investigated the association of serum copper concentration with the risk of acute myocardial infarction in randomly selected men aged 42, 48, 54, or 60 years who had no symptomatic ischemic heart disease at entry. Baseline examinations in the Kuopio Ischemic Heart Disease Risk Factor Study in Eastern Finland were done during 1984–1988. In Cox multivariate survival models adjusting for age, examination year, ischemic electrocardiogram in exercise, maximal oxygen uptake, diabetes, family history of ischemic heart disease, cigarette-years, mean systolic blood pressure, serum HDL-cholesterol subfraction HDL2 and LDL-cholesterol concentrations and blood leukocyte count, serum copper concentration in the two highest tertiles (1.02–1.16 mg/L and 1.17 mg/L or more) associated with 3.5-fold (95 % confidence interval (CI) 1.3–9.4, $P < 0.05$) and 4.0-fold (95 % CI 1.5–10.8, $P < 0.01$) risk of acute myocardial infarction. These data indicated that high copper status, reflected by elevated serum copper concentration, is an independent risk factor for ischemic heart disease.

Jain and Mohan (1991) determined serum copper and zinc estimations in humans to find their diagnostic and prognostic value in cases of myocardial infarction. Following infarction, there was an increase in serum copper levels from the first 24 h up to the 7th day, with gradual decline that did not reach the normal value up to the 14th day. The serum zinc levels declined in the first 24 h until the 4th day and increased to the normal value on the 14th day. It is concluded that, for diagnosis of myocardial infarction, serum zinc levels are more useful during the first week and copper levels in the second week after the onset of infarction.

Walter et al. (1991) evaluated copper, zinc, manganese, magnesium, and other indices of peroxidative status in diabetic and nondiabetic human subjects. Hyperzincuria and hypermagnesuria were evident in diabetic subjects compared with control subjects. There were no differences in plasma magnesium or whole-blood manganese between groups. Plasma copper was higher and plasma zinc was lower in diabetic than in control subjects. Data were viewed with respect to specific diabetes-associated complications; diabetic subjects with retinopathy, hypertension, or microvascular disease had higher plasma copper concentrations compared with both diabetic subjects without complications and with control subjects. There were no significant differences between control and diabetic subjects in erythrocyte copper–zinc superoxide dismutase activity or whole-blood glutathione peroxidase or glutathione reductase activities. Plasma peroxide concentrations were higher in diabetic than control subjects. They concluded that diabetes can alter copper, zinc, magnesium, and lipid peroxidation status. Perturbations in mineral metabolism are more pronounced in diabetic populations with specific complications. It is not known whether differences in trace element status are a consequence of diabetes, or alternatively, whether they contribute to the expression of the disease.

Swanson et al. (1990) reported that selenium intake (micrograms/kg body weight) was strongly correlated (all values, $P < 0.01$) with selenium concentration of serum ($r = 0.63$), whole blood ($r = 0.62$), and toenails ($r = 0.59$). Men and women had similar mean values of serum, whole blood, and toenail selenium despite higher selenium intakes in men. Smokers had lower tissue selenium concentrations than did nonsmokers due, at least in part, to lower selenium intake. Age was not associated with tissue selenium content. Of the variables examined, selenium intake was clearly the strongest predictor of tissue selenium concentration.

Fischer et al. (1990) determined the effects of zinc supplementation on the copper status of healthy adult men, as assessed by the activities of the copper metalloenzymes, plasma ferroxidase (ceruloplasmin), and erythrocyte Cu–Zn superoxide dismutase. No significant differences in the plasma copper levels or the ferroxidase activities between the supplemented and control groups could be detected at 2, 4, or 6 weeks. Plasma zinc increased and erythrocyte Cu–Zn superoxide dismutase decreased in the supplemented group, the difference between the groups becoming significant at 6 weeks ($P < 0.05$). This suggested that the zinc supplements decreased the copper status of the experimental group.

Goto et al. (1990) studied that magnesium (Mg) deficiency was present in patients with variant angina; 24-h Mg retention of low-dose Mg (0.2 mEq/kg lean body weight) administered intravenously over 4 h in 20 patients with variant angina was examined. No patient had received calcium antagonists before or during the study. The mean serum Mg concentrations in the patients with variant angina and the control subjects were 2.1 ± 0.05 and 2.1 ± 0.03 mg/dL, respectively (difference not significant). However, 24-h Mg retention in the patients with variant angina was 60 ± 5 %, while that in the control subjects was 36 ± 3 % ($P < 0.001$), suggesting that Mg deficiency is present in at least some patients with variant angina. The mean serum Mg concentrations before and after calcium antagonist treatment in patients with variant angina were 2.1 ± 0.09 and 2.1 ± 0.07 mg/dL, respectively (difference not significant). However, 24-h Mg retention decreased significantly ($P < 0.01$) from 60 ± 6 to 34 ± 7 % after the treatment. They concluded that there is Mg deficiency in many patients with variant angina and it is corrected after treatment with calcium antagonists.

Klevay (1989) observed that atherosclerosis is a complex process beginning early in life and often leading to death from ischemic heart disease in middle age. Hundreds of factors are said to contribute to this risk. More than 50 similarities between animals deficient in copper and people with ischemic heart disease have been identified. Some of the more important characteristics of this illness have been produced in experiments in which men and women were fed diets low in copper.

Diets with similarly low amounts of copper are readily available to the population at large. More aspects of the anatomy, biochemistry, chemistry, epidemiology, pathogenesis, and pathophysiology of ischemic heart disease can be explained by considering this illness to be a problem of copper deficiency than by considering any of several other explanations that have been offered.

Mocchegianai et al. (1989) reported that zinc is required for optimal functioning of the immune system. It was recently reported that one of the best-known thymic hormones responsible for the maturation and differentiation of the thymus-derived T-lymphocyte line, that is, serum thymic factor (STF), is biologically active only when bound to zinc ions; in this form, it has been called thymulin (Zn-STF). Because low serum and tissue zinc values have been reported to occur in diabetic conditions, and because defects of T-lymphocyte-dependent functions are also present in diabetic patients, even metabolically well-controlled diabetic patients, they investigated the serum level of zinc and the plasma level of both active Zn-STF and inactive STF thymic hormones in young patients suffering from type I (insulin-dependent) diabetes. Serum zinc levels were significantly reduced in diabetic conditions and did not correlate with the degree of metabolic compensation measured by glycosylated hemoglobin. In diabetes, the active form of thymulin was strongly reduced, whereas the inactive form was abnormally elevated. In vitro zinc addition to diabetic plasma samples also induced zinc saturation of inactive thymic hormone molecules: the total thymic hormone measured in these experimental conditions showed values in diabetic patients comparable with those observed in healthy age-matched individuals, suggesting that low thymulin levels recorded in diabetic conditions are due not to a thymic failure in synthesizing and secreting thymic hormone but to a peripheral defect in zinc saturation of the hormone molecules. The zinc-dependent failure of thymic hormone, present even in fairly compensated diabetic conditions, might account for the apparent insulin-independent immunological abnormalities associated with type I diabetes.

Harlan (1988) reported that numerous observations have indicated a relationship between moderate or heavy lead exposure and high blood pressure. They analyzed data from the National Health and Nutrition Examination Survey II for persons 12–74 years of age. Significant correlations were found between blood lead and blood pressure for each race–gender group, and blood lead levels were significantly higher in groups with high diastolic blood pressure (greater than 90 mmHg). Multiple stepwise regression models were developed to predict blood pressure. After adjusting for age, race, and body mass index, blood lead levels were significantly related to systolic and diastolic pressures in males but not in females. These findings and those from other studies confirmed the relationship of blood lead and blood pressure at relatively low levels commonly observed in the general population. The strength and importance of this relationship require further study through epidemiological and metabolic investigations.

Fujimoto (1987) determined the relationships between blood trace metal concentrations and the clinical status of patients with cerebrovascular disease, gastric cancer, and diabetes mellitus. Patients with cerebrovascular disease showed generally lower concentrations than normal subjects, while the gender difference of the blood metal concentrations showed a pattern similar to that of normal subjects. In some combination, significant correlations were observed between blood metal concentrations and clinical biochemical parameters. Plasma copper concentrations had a significant positive correlation with catalase. The blood copper concentration of patients with diabetes mellitus showed a distribution pattern similar to that of healthy subjects. Therefore, copper is not considered to be an important factor in diabetes mellitus. Diabetic patients treated by insulin injection showed increased blood zinc concentrations. Chromium, which is contained in GTF (glucose tolerance factor), showed lower blood concentrations in patients with severe complications, such as retinopathy or nephropathy. Therefore, it appears that chromium plays an important role in advancing diabetes mellitus.

Tiber et al. (1986) suggested an association between dietary copper deficiencies, alone or with an attendant elevated intake of zinc, and an increase in serum total cholesterol. These findings led some to theorize that a dietary imbalance of zinc and copper may be a factor in the etiology of coronary heart disease. Plasma zinc and copper levels were measured in adult male patients with confirmed CAD and compared with serum levels of lipids and lipoproteins in the same patients. The results were compared to the accepted values in normal adult males. In subjects with significant coronary artery atherosclerosis (greater than or equal to 50 % luminal occlusion), there was no correlation between plasma zinc and copper with serum lipids or lipoproteins. However, total cholesterol was significantly correlated with low-density lipoprotein (LDL) cholesterol and triglycerides and inversely correlated with HDL-cholesterol/TC ratio. Although dietary zinc and/or copper may influence the plasma levels of these trace metals, they showed that there was no association between plasma zinc or copper and the serum levels of lipids or lipoproteins; this indicated that these trace metals are of doubtful value as markers for coronary atherosclerosis.

Bales et al. (1986) measured plasma uptake of oral dose of zinc at 0, 1, 2, 3, and 4 h post-dose in elderly (mean age = 7.25 years) and young (mean age = 24.0 years) subjects selected from a group of healthy nonsmokers. Elderly and young subjects were divided into two groups based upon low or high detection thresholds for solutions of sodium chloride and sucrose. Mean fasting concentrations of plasma zinc were almost identical for the two age groups. Plasma response to a zinc dose was lower ($P < 0.05$) in the elderly compared to that in the young, as indicated by plasma zinc levels at hours 2 and 3 post-dose and by total area under the response curve. Moderate but significant age-associated increases in detection thresholds for both sodium chloride and sucrose were found; taste thresholds were not correlated to plasma uptake of the zinc dose in either age group.

Kromhout et al. (1985) determined trace metals and coronary heart disease risk indicators among men aged 57–76 years in the town of Zutphen, the Netherlands. After uni- and multivariate regression analysis, the following statistically significant relations were found: serum zinc was inversely related to resting heart rate; serum copper was positively related to cigarette smoking and inversely to HDL cholesterol; blood cadmium was strongly positively related to cigarette smoking and inversely to Quetelet index; the positive relation between blood lead and cigarette smoking was of borderline significance; and blood lead was related to blood pressure, with the relation being stronger for systolic than for diastolic blood pressure.

Singh et al. (1985) carried out serum copper estimations in cases of AMI, cases of angina, and age- and sex-matched healthy controls. A highly significant degree of rise in serum copper levels was observed in patients with AMI as compared to cases of angina and controls. The levels showed a gradual rise with peak on 7th day followed by a gradual decline returning to normal on 28th day. The pattern was the same both in complicated and uncomplicated cases except that values were still higher on the 28th day in complicated cases of acute myocardial infarction. Mean peak serum copper levels were significantly higher ($P < 0.001$) in complicated cases of acute myocardial infarction as compared to uncomplicated cases. Significant correlation was found between serial serum copper changes and creatinine phosphokinase and lactic dehydrogenase levels.

Altura and Altura (1985) observed that hypomagnesemia was a rare entity in clinical practice. It is clear, however, from newer studies that the overall incidence of hypomagnesemia in hospitalized patients can range from 7 to 52 %. The greatest association of hypomagnesemia in hospitalized patients appeared to be in hypokalemic states and in patients confined to intensive care units. Most of these patients demonstrated cardiovascular abnormalities, ranging from cardiac arrhythmias and atrial fibrillation to hypertension. On the basis of primarily epidemiological and experimental findings, it has been suggested that there may be a strong association between the dietary intake of Mg (and errors in the Mg metabolism and distribution of Mg in the body), the concentration of this element in the myocardium and blood vessels, and the risk for development of cardiac arrhythmias, sudden death ischemic heart disease, hypertension, transient ischemic attacks, strokes, and preeclampsia–eclampsia. Clinical trials utilizing Mg as a therapeutic tool to treat refractory arrhythmias, digitalis toxicity-associated arrhythmias, myocardial infarctions, diabetic angiopathy, transient ischemic attacks, cerebral resuscitation, hypertension, and "classical" migraine are under way and to an extent have been successful. Careful assessment of serum, blood cells, and urine for free versus bound Mg should be done routinely in cardiovascular disease and high-risk patients.

Cappuccio et al. (1985) reported 17 unselected patients with mild to moderate essential hypertension and whose average supine blood pressure after 2 months' observation with no treatment was 154/100 mmHg were entered into a double-blind randomized crossover study of 1 month's treatment with magnesium aspartate (15 mmol magnesium/day) and treatment with placebo for a further month. This preparation of magnesium was well tolerated and did not cause diarrhea. Despite a significant increase in plasma magnesium concentration and a significant increase in urinary excretion of magnesium while taking magnesium aspartate, there was no fall in blood pressure compared with either treatment with placebo or values before treatment. The results provided no evidence for a role of dietary magnesium in the regulation of high blood pressure and are contrary to recent speculations.

Heinecke et al. (1984) reported that modification of low-density lipoproteins by human arterial smooth muscle cells were characterized by increased electrophoretic mobility and increased content of malondialdehyde-like oxidation products reactive with thiobarbituric acid. Lipoprotein modification was promoted by micromolar concentrations of iron or copper in the culture medium and was metal ion concentration- and time-dependent. The ability of diverse media to promote smooth muscle cell-mediated low-density lipoprotein modification correlated with their iron concentration. Therefore, metal ion concentration of culture media contributes substantially to low-density lipoprotein modification in vitro. Human monocyte-derived macrophages took up and esterified the cholesterol from modified low-density lipoprotein more extensively than from native low-density lipoprotein. Metal ion-mediated modification of low-density lipoprotein may be a contributing factor to the pathogenesis of arteriosclerosis.

Uza et al. (1984) determined serum concentration of Na, K, Ca, Mg, and inorganic phosphate as well as serum levels of Zn and Cu in control subjects and in patients with essential arterial hypertension (EAH) divided according to the stage of the disease. No significant differences were found between the serum mean levels of Na, K, Ca, Mg, Zn, and Cu in controls and in patients with EAH. A significant decrease of the serum Zn was noted in the third stage of EAH. A number of cases with hypomagnesemia and/or hypopotassemia probably caused by a long-term uncontrolled therapy were also detected. The concentration of inorganic phosphate was significantly lower in patients with EAH associated with overweight than in hypertensive patients with normal body weight and in controls. It was considered that a sustained study of the complex interrelationship between electrolyte interaction and the functional aspects of the arterial wall could still contribute to a better understanding of pathogenic aspects of EAH and of its complications including those subsequent to modern diuretic therapy.

Durlach and Collery (1984) studied that diabetes mellitus is the most common pathological state in which secondary magnesium deficiency occurs. Magnesium metabolism abnormalities vary according to the multiple clinical forms of diabetes: plasma magnesium is more often decreased than red blood cell magnesium. Plasma Mg levels were correlated mainly with the severity of the diabetic state, glucose disposal, and endogenous insulin secretion. Various mechanisms are involved in the induction of Mg depletion in diabetes mellitus, that is, insulin and epinephrine secretion; modifications of the vitamin D metabolism; decrease of blood P, vitamin B6, and taurine levels; increase of vitamin B5, C, and glutathione turnover; and treatment with high levels of insulin. K depletion in diabetes mellitus is well known. Some of its mechanisms are concomitant to those of Mg depletion. But their hierarchic importance is not the same: that is, insulin hyposecretion is more important versus K^+ than versus Mg^{2+}. Insulin increases the cellular inflow of K^+ more than that of Mg^{2+} because there is more free K^+ (87 %) than Mg^{2+} (30 %) in the cell. The consequences of the double Mg–K depletion are either antagonistic, that is, versus insulin secretion (increased by K^+, decreased by Mg^{2+}), or agonistic, that is, on the membrane (i.e., Na^+K^+ATPase), tolerance of glucose oral load, renal disturbances. The real importance of these disorders in the diabetic condition is still poorly understood. Retinopathy and microangiopathy are correlated with the drop of plasma and red blood cell Mg. K deficiency increases the noxious cardiorenal effects of Mg deficiency. The treatment should primarily insure diabetic control.

Canfield et al. (1984) reported that zinc concentrations in plasma, hair, and urine from children and young adults with insulin-requiring or type I diabetes mellitus were significantly correlated with height, weight, and age, as well as with indices of metabolic control, that is, fasting serum glucose, percent glycosylated hemoglobin (HbA1), and 24-h urine glucose and insulin excretion. Urinary zinc excretion was greater in subjects than in controls ($P < 0.0001$) and significantly correlated with urine glucose ($P < 0.004$, $r = 0.35$) and volume ($P < 0.0007$, $r = 0.40$). Urinary zinc and volume were not correlated in controls. Hyperzincuria in the subjects was not secondary to hyperinsulinuria, although zinc and insulin excretion was significantly correlated in controls ($P < 0.03$, $r = 0.63$). Zinc in insulin preparations could not explain the excessive zinc excretion. Mean fasting plasma zinc was significantly higher than in controls and positively correlated with height for age, while being inversely correlated with age, duration of diabetes, HbA1, urine volume, and glucose excretion. Both the mean and range of hair zinc concentration in the subjects were not different from controls. Male subjects with diabetes had a significantly lower hair concentration when compared with female subjects with diabetes ($P < 0.0009$). Zinc homeostasis appears to be altered as a consequence of glucose intolerance in diabetes. Continued urinary zinc losses over time may result in a zinc deficiency state not demonstrable by altered zinc concentrations in plasma and hair.

Noto et al. (1983) reported that in diabetics, even in those without complications and/or alterations of lipid metabolism, higher levels of cupremia were found than in the controls; these increased levels were not correlated with the duration of the disease, and the cupremia was more evident in the older patients and in those with complications. However, the increases were not dissimilar from those found in nondiabetic arteriosclerosis. Their findings favored the hypothesis that after some years and by affecting fatty acid metabolism, higher levels of cupremia enhance the appearance of diabetic vasculopathy.

Hurley et al. (1983) summarized the origins of nutritional trace element deficiencies. Inadequate intake results in primary deficiency, whereas secondary or conditioned deficiencies can arise in several ways including trace element interactions. Evidence is presented and discussed for interactions of essential trace elements during prenatal and early postnatal development. Analysis of fetal outcome and copper and zinc concentrations of maternal and fetal livers showed that although there is an interaction between these metals, it occurs only at levels of dietary copper deficiency. Iron and manganese interact so that high levels of one depress absorption of the other. Mice fed iron-supplemented diets had liver manganese concentrations lower than those of unsupplemented mice. Iron supplements at high but not low levels also depressed absorption of zinc. Conversely, zinc deficiency in pregnant rats caused higher than normal concentrations of iron in maternal and fetal liver. Trace element analyses of proprietary infant formulas indicate that in some, concentrations and ratios of these trace elements may be incorrect. The effects of essential trace element interactions during development should be further investigated. Caution is urged in considering levels of trace element supplements during pregnancy, lactation, or early childhood.

Kinlaw et al. (1983) investigated zinc metabolism in patients with stable type II diabetes mellitus. Twenty-five percent of these patients had depressed serum zinc concentrations, and all demonstrated hyperzincuria. Urinary zinc loss was greater when proteinuria was present and correlated with the mean serum glucose concentration. Studies of gastrointestinal zinc absorption suggested zinc malabsorption in patients with type II diabetes mellitus. Glucose infusion in normal dogs produced hyperzincuria without a diminution in serum zinc. They concluded that hyperzincuria, resulting from a glucose-mediated process that is not osmotic, interacts with impaired zinc absorption to produce zinc deficiency in patients with type II diabetes mellitus.

Levine et al. (1983) observed a structural and functional relationship between zinc and insulin. Zinc concentrations of various tissues from genetically diabetic and streptozotocin-induced diabetic mice and their appropriate control mice were determined. The zinc concentrations were depressed in serum and femur of C57BL/Ks-db+/db+ mice (db/db) when compared with their nondiabetic heterozygote controls (db/m) and homozygous controls (m/m). No differences were noted in the hepatic or renal Zn concentration of the db/db, db/m, or m/m mice. Zinc supplementation in the drinking water for a 4-week period had no effect on serum or tissue zinc concentration. Hyperzincuria was noted in the db/db mice. No differences were noted in the Zn concentration of serum or tissue in streptozotocin-induced diabetic mice compared to their controls. These data suggest that zinc deficiency may play a role in the pathogenesis of the insulin resistance present in type II (insulin-independent) diabetics.

McCarron (1982) reported that primary disturbances of divalent ion metabolism contribute to the development and maintenance of hypertension. Representative interactions of calcium, magnesium, and phosphorus with normal cardiovascular physiology were presented. Established and postulated abnormalities of divalent ion metabolism associated with human and experimental hypertension were reviewed. The influence of calcium balance on blood pressure development in the young spontaneously hypertensive rat was demonstrated by the results of a diet intervention study. The SHRs' blood pressures stratified inversely ($P < 0.001$) based upon the calcium content. The low-calcium animals experienced a more rapid and greater rise in blood pressure between 4 and 20 weeks of age ($P < 0.01$). Blood pressures of the supplemented SHRs (4 %) peaked at a lower value (174 vs. 192 mmHg, $P < 0.01$). After maturity, the 4 % SHRs experienced attenuation ($P < 0.01$) of their hypertension (154 $\pm$ 7 mmHg, 4 % SHR vs. 176 $\pm$ 7 mmHg, 0.5 % SHR). It was proposed that membrane-associated bioavailable Ca^{2+} is reduced in the SHR and possibly in human hypertension. Dietary calcium supplementation may reverse this defect, resulting in cell membrane stabilization and vascular smooth muscle relaxation.

McNair et al. (1982) studied interrelations between magnesium and glucose metabolism in insulin-treated diabetic outpatients aged 7–70 years. All had normal serum creatinine concentrations (below 115 μmol/L), and none had other diseases or received drugs known to interfere with mineral metabolism. A definite hypomagnesemia (< normal mean −2 SD) and hypermagnesiuria (> normal mean +2 SD) occurred in 38.6 and 55 % of the patients. In the presence of hypermagnesiuria, the serum magnesium concentration was inversely correlated to the urinary magnesium excretion rate ($r = -0.23$, $P < 0.02$). Serum magnesium correlated inversely with both fasting blood glucose ($r = -0.32$, $P < 0.001$) and the urinary glucose excretion rate ($r = -0.022$, $P < 0.005$). The urinary magnesium excretion rate correlated directly with the same variables ($r = 0.27$, $P < 0.001$ and $R = 0.58$, $P < 0.001$, respectively). These data indicated that the net tubular reabsorption of magnesium is decreased in diabetic patients in presence of hyperglycemia, leading to hypermagnesuria and hypomagnesemia.

Helgeland et al. (1982) collected serum samples from the adult population, age groups from 20 to 54 years, in different Norwegian municipalities and analyzed for zinc and copper by atomic absorption spectroscopy. Significant differences were found between several of the municipalities when the mean concentration of zinc in serum in 200 randomized samples was compared, with only two municipalities being different for copper. The values for zinc ranged from 13.8 to 18.3 μmol/L, and copper varied between 16.3 and 19.2 μmol/L. An age-related increase in the copper concentration was evident in the male population, and age-adjusted means showed a slight but significantly higher serum copper concentration in females (18.4 μmol/L) than in males (16.5 μmol/L). For zinc, the opposite sex relationship was indicated with the highest values in males, 15.8 compared to 15.1 μmol/L in serum from females. No significant correlations were found between the concentrations of zinc and copper in serum. In all age groups of women, however, a small negative correlation was found giving a significant tendency.

Levin et al. (1981) assayed leukocyte and erythrocyte magnesium in healthy subjects and insulin-dependent diabetic patients. Plasma magnesium concentration (mean ± standard error of mean) was significantly lower in the diabetic patients (0.80 ± 0.02 mmol/L), compared with the healthy subjects (0.90 ± 0.02 mmol/L, $P < 0.001$), but the leukocyte and erythrocyte magnesium content was not significantly different in the diabetic patients (34.5 ± 0.8 and 6.2 ± 0.2 mmol/kg dry solids) compared with the healthy subjects (35.5 ± 0.8 and 6.5 ± 0.11 mmol/kg dry solids). Plasma magnesium in diabetic patients (0.74 ± 0.01 mmol/L) was significantly lower than in healthy subjects (0.85 ± 0.02 mmol/L, $P < 0.001$), but there was no significant difference in the mean muscle magnesium content (43.0 ± 0.7 compared with 40.7 ± 0.9 mmol/kg dry solids in the diabetic patients).

Bettger and O'Dell (1981) reported large number of recognized zinc metalloenzymes; the activities of only a few were significantly decreased in severely zinc-deficient animals. On the other hand, physiological pathology was manifested rapidly after dietary zinc deprivation. This showed that zinc exerts physiological and biochemical roles other than as a component of the known zinc metalloenzymes. The research results reviewed here suggested that zinc plays an important role in the maintenance of membrane structure and function.

Mather et al. (1979) measured plasma magnesium (Mg) concentrations of unselected diabetic outpatients and control subjects by atomic absorption spectrophotometer. Mean plasma Mg (± SD) was significantly lower in the diabetic patients (0.737 ± 0.071 mmol/L) than in the control subjects (0.810 ± 0.057 mmol/L), and (25 %) diabetics had values below those of all control subjects except one. Plasma Mg correlated best with clinical blood glucose concentration ($r = -0.32$, $P < 0.001$), and other significant associations were observed with glycosuria, age, sex, insulin therapy, and biguanide therapy. Although its significance is unclear, hypomagnesemia could conceivably predispose to ischemic heart disease in diabetes.

Frieden and Hsieh (1976) surveyed the chemistry and state copper in the molecule because of the implications of the recent data of Ryden. Ceruloplasmin, the blue copper protein of vertebrate plasma, has been reviewed mainly from a functional point of view. They suggested that unless special precautions are taken in the isolation of ceruloplasmin, i.e. degradation, probably proteolytic, fragments of various sizes are produced. Three types of copper atoms have been identified in ceruloplasmin, but their amino acid environment was still unknown. Ceruloplasmin possesses significant oxidase activity toward Fe (II) and numerous aromatic amines and phenols. Its ferroxidase activity has led to the discovery that it is a molecular link between copper and iron metabolism. Ceruloplasmin mobilizes iron into the plasma from iron storage cells in the liver. An equally important duty is that ceruloplasmin, after its rapid biosynthesis in the liver, serves as a major copper transport vehicle, comparable to transferrin. Evidence was accumulating that the copper atoms of ceruloplasmin are a prerequisite for copper utilization in the biosynthesis of cytochrome oxidase and other copper proteins. The ability of ceruloplasmin to release copper at specific cellular sites may be related to its broad substrate spectrum of biological reducing agents. A possible third role of ceruloplasmin is as a contributor to the regulation of the balance of biogenic amines through its oxidase action on the epinephrine and the hydroxyindole series. Thus, ceruloplasmin is a copper protein with several important functions, all of which are directly related to its oxidase activity.

Klevay (1975) reported that metabolic imbalance in regard to zinc and copper is a major factor in the etiology of CHD. This metabolic imbalance is either a relative or an absolute deficiency of copper characterized by a high ratio of zinc to copper. The imbalance resulted in hypercholesterolemia and increased mortality due to coronary heart disease. The imbalance can occur due to the amounts of zinc and copper in human food, to lack of protective substances in food or drinking water, and to alterations in physiological status that produce adverse changes in the distribution of zinc and copper in certain important organs. Because no other agent, with the possible exception of cholesterol, has been related so closely to task, the ratio of zinc to copper may be the preponderant factor in the etiology of coronary heart disease.

Handjani et al. (1974) examined the serum zinc concentration in patients, some of them with well-documented myocardial infarction and a few patients with chest pain caused by myocardial ischemia without infarction. The zinc level fell sharply in the myocardial infarction patients within a day of onset and then rose to normal values within 7–10 days. There was no comparable change from normal values in the serum zinc level in seven patients with myocardial ischemia, although two showed bordered values. It was suggested that under appropriate conditions, measurement of serum zinc concentration levels may be a useful aid for differentiation of myocardial ischemia from infarction.

Rotruck et al. (1973) observed that when hemolyzates from erythrocytes of selenium-deficient rats were incubated in vitro in the presence of ascorbate or H_2O_2, added glutathione failed to protect the hemoglobin from oxidative damage. This occurred because the erythrocytes were practically devoid of glutathione peroxidase activity. Extensively purified preparations of glutathione peroxidase contained a large part of the Se of erythrocytes labeled in vivo. Many of the nutritional effects of selenium can be explained by its role in glutathione peroxidase.

Halsted and Smith (1970) measured plasma zinc levels by atomic absorption spectrophotometer in healthy adults and children, in patients with a variety of diseases, in pregnancy, and in women taking oral contraceptives. The mean level was 96 μg/100 mL for healthy adults and 89 μg/100 mL for healthy children, and the concentration was constant with no significant variation attributable to sex, food consumption, or diurnal variation. Abnormally low values were obtained in alcoholic cirrhosis, other types of liver disease, active tuberculosis, indolent ulcers, uremia, before and after a single hemodialysis, myocardial infarction, nontuberculous pulmonary infection, Down's syndrome, cystic fibrosis with growth retardation, growth-retarded Iranian villagers, pregnancy, and in women taking oral contraceptives. In cystic fibrosis without growth retardation and in inactive tuberculosis, there was no significant decrease. No conditions have been observed with a higher-than-normal plasma zinc concentration.

References

Akyüz F, Önder E, Erden M (1993) Evaluation of serum magnesium, zinc, copper and ascorbic acid levels in patients with hypertension and atherosclerotic heart diseases. Turk J Med Res 11 (6):273–276

Al-Saleh E, Nandakumaran M, Al-Shammari M, Makhseed M, Sadan T, Harouny A (2005) Maternal-fetal status of copper, iron, molybdenum, selenium and zinc in insulin-dependent diabetic pregnancies. Arch Gynecol Obstet 271(3):212–217

Altura BM, Altura BT (1985) New perspectives on the role of magnesium in the pathophysiology of the cardiovascular system. Magnesium 4:226–244

Ascherio A, Hennekens C, Willett WC, Sacks F, Rosner B, Manson J, Witteman J, Stampfer MJ (1996) Prospective study of nutritional factors, blood pressure and hypertension among US women. Hypertension 27(5):1065–1072

Bales CW, Steinman LC, Freeland-Graves JH, Stone JM, Young RK (1986) The effect of age on plasma zinc uptake and taste acuity. Am J Clin Nutr 44:664–669

Bettger W, O'Dell B (1981) A critical physiological role of zinc in the structure and function of biomembranes. Life Sci 28:1425–1438

Canfield WK, Hambidge KM, Johnson LK (1984) Zinc nutriture in type I diabetes mellitus: relationship to growth measures and metabolic control. J Pediatr Gastroenterol Nutr 3 (4):577–584

Cappuccio FP, Markandu ND, Beynon GW, Shore AC, Sampson B, MacGregor GA (1985) Lack of effect of oral magnesium on high blood pressure: a double blind study. Br Med J 291:235–238

Choi DW (1998) Zinc and brain injury. Ann Rev Neurosci 21:347–375

Cooper GJ, Chan YK, Dissanayake AM, Leahy FE, Keogh GF, Frampton CM, Gamble GD, Brunton DH, Baker JR, Poppitt SD (2005) Demonstration of a hyperglycemia-driven pathogenic abnormality of Copper homeostasis in diabetes and its reversibility by selective chelation: quantitative comparisons between the biology of copper and eight other nutritionally essential elements in normal and diabetic individuals. Diabetes 54(5):1468–1476

Durlach J, Collery P (1984) Magnesium and potassium in diabetes and carbohydrate metabolism. Magnesium 3:315–323

Ekin S, Mert N, Gunduz H, Meral I (2003) Serum sialic acid levels and selected mineral status in patients with type 2 diabetes mellitus. Biol Trace Elem Res 94(3):193–201

Elwood PC, Pickering J (2002) Magnesium and cardiovascular disease: a review of epidemiological evidence. J Clin Basic Cardiol 5:61–66

Ferrara LA, Iannuzzi R, Castaldo A, Iannuzzi A, Dello Russo A, Mancini M (1992) Long-term magnesium supplementation in essential hypertension. Cardiology 81:25–33

Fischer PWF, L'Abbé MR, Giroux A (1990) Effects of age, smoking, drinking, exercise and estrogen use on indices of copper status in healthy adults. Nutr Res 10:1081–1090

Ford ES (1999) Serum magnesium and ischemic heart disease: findings from a national sample of US adults. Int J Epidemiol 28:645–651

Frieden E, Hsieh HS (1976) Ceruloplasmin: the copper transport protein with essential oxidase activity. Adv Enzymol Relat Areas Mol Biol 44:187–236

Fujimoto S (1987) Studies on the relationship between blood trace metal concentration and the clinical status of patients with cerebrovascular disease, gastric cancer and diabetes mellitus. Hokoido Igaku Zasshi 62:913–932

Geleinjoise JM, Witteman JC, den Breeijen JH, Hofman A, de Jong PT, Pols HA, Grobbee DE (1996) Dietary electrolyte intake and blood pressure in older subjects: the Rotterdam Study. J Hypertens 14(6):737–741

Goto K, Yasue H, Okumura K, Matsuyama K, Kugiyama K, Miyagi H, Higashi T (1990) Magnesium deficiency detected by intravenous loading test in variant angina pectoris. Am J Cardiol 65:709–712

Halsted JA, Smith JC (1970) Plasma-zinc in health and disease. Lancet 1(7642):322–324

Handjani AM, Smith JC Jr, Herrmann JB, Halsted JA (1974) Serum zinc concentration in acute myocardial infarction. Chest 65:185

Harlan WR (1988) The relationship of blood lead levels to blood pressure in the US population. Environ Health Perspect 78:9–13

Heinecke JW, Rosen H, Chait A (1984) Iron and copper promote modification of low density lipoprotein by human arterial smooth muscle cells in culture. J Clin Invest 74:1890–1894

Helgeland K, Haider T, Johnsan J (1982) copper and zinc in human serum in Norway. Relationship to geography, sex and age. Scand J Clin Lab Invest 42(1):35–39

Hennig B, Toborek M, McClain CJ (1996) Antiatherogenic properties of zinc: implications in endothelial cell metabolism. Nutrition 12(10):711–717

Hennig B, Meerarani P, Toborek M, McClain CJ (1999) Antioxidant-like properties of zinc in activated endothelial cells. J Am Coll Nutr 18:152–158

Holecek V, Racek J, Jerabek Z (1995) Administration of multivitamin combinations and trace elements in diabetes. Cas Lek Cesk 134:80–83

Huang KC, Lin WY, Lee LT (2002) Four anthropometric indices and cardiovascular risk factors in Taiwan. Int J Obes Relat Metab Disord 26:1060–1068

Hurley LS, Keen CL, Lonnerdal B (1983) Aspects of trace element interactions during development. Fed Proc 42:1735–1739

Institute of Medicine (2002) Dietary reference intakes: vitamin A, vitamin K, arsenic, boron, chromium, copper, iodine, iron, manganese, molybdenum, nickel, silicon, vanadium, and zinc. Food and Nutrition Board. National Academy Press, Washington, DC, pp 224–257

Jain VK, Mohan G (1991) Serum zinc and copper in myocardial infarction. Biol Trace Elem Res 31:317

Kazemi-Bajestani SM, Ghayour-Mobarhan M, Ebrahimi M, Moohebati M, Esmaeili HA, Parizadeh MR, Aghacizadeh R, Ferns GA (2007) Serum copper and zinc concentrations are lower in Iranian patients with angiographically defined coronary artery disease than in subjects with a normal angiogram. J Trace Elem Med Biol 21:22–28

Kazi TG, Afridi HI, Kazi N, Jamali MK, Arain MB, Sarfraz RA, Jalbani N, Ansari R, Shah AQ, Memon AU, Khandhro GA (2008) Distribution of zinc, copper and iron in biological samples of Pakistani myocardial infarction (1st, 2nd and 3rd heart attack) patients and controls. Clin Chim Acta 389(1–2):114–119

Kinlaw WB, Levine AS, Morley JE, Silvis SE, McClain CJ (1983) Abnormal zinc metabolism in type II diabetes mellitus. Am J Med 75:273–277

Klevay LM (1975) Coronary heart disease: the zinc/copper hypothesis. Am J Clin Nutr 28:764–774

Klevay LM (1989) Ischemic heart disease as copper deficiency. Adv Exp Med Biol 258:197–208

Kosar F, Taskapan C, Kucukbay Z (2007) Serum levels of selenium, zinc and copper in patients with coronary artery ectasia. Indian Heart J 59(1):38–41

Kromhout D, Wibowo AAE, Herber RFM (1985) Trace metals and coronary heart disease risk indicators in 152 elderly men (The Zutphen Study). Am J Epidemiol 122:378–385

Kruse-Jarres JD, Rükgauer M (2000) Trace elements in diabetes mellitus. Peculiarities and clinical validity of determinations in blood cells. J Trace Elem Med Biol 14(1):21–27

Landmark K, Urdal P (1993) Serum magnesium and potassium in acute myocardial infarction: relationship to existing beta-blockade and infarct size. Angiology 44:347–352

Levin GE, Mather HM, Pilkington TRE (1981) Tissue magnesium status in diabetes mellitus. Diabetologia 21:131–134

Levine AS, McClain CJ, Handwerger BS, Brown DM, Morley JE (1983) Tissue zinc status of genetically diabetic and streptozotocin-induced diabetic mice. Am J Clin Nutr 37 (3):382–386

Liao F, Folsom A, Brancati F (1998) Is low magnesium concentration a risk factor for coronary heart disease? The Atherosclerosis Risk in Communities (ARIC) Study. Am Heart J 136:480–490

Linder MC, Hazegh-Azam M (1996) Copper biochemistry and molecular biology. Am J Clin Nutr 63:797S–811S

Ma J, Betts NM (2000) Zinc and copper intakes and their major food sources for older adults in the 1994-96 continuing survey of food intakes by individuals (CSF 11). J Nutr 130:2838–2843

Maj-Zurawska M (1994) Clinical findings on human blood with the KONE ISE for Mg. Scand J Clin Lab Invest 54(suppl 217):69–76

Martin-Lagos F, Navarro-Alarcon M, Terres-Martos C, Lopez G, de la Serena H, Lopez-Martinez MC (1997) Serum copper and zinc concentrations in serum from patients with cancer and cardiovascular disease. Sci Total Environ 204:27–35

Mather HM, Nisbet JA, Burton GH (1979) Hypomagnesaemia in diabetes. Chim Clin Acta 95:235–242

McCarron DA (1982) Calcium, magnesium, and phosphorus balance in human and experimental hypertension. Hypertension 4(Suppl III):III-27–III-33

McMaster D, McCrum E, Patterson CC, Kerr MM, O'Reilly D, Evans AE (1992) Serum copper and zinc in random samples of the population of Northern Ireland. Am J Clin Nutr 56:440–446

McNair P, Christensen MS, Christiansen C, Madsbad S, Transbol I (1982) Renal hypomagnesaemia in human diabetes mellitus: its relation to glucose homeostasis. Eur J Clin Invest 12:81–85

Meerarani P, Ramadass P, Toborek M, Bauer HC, Bauer H, Hennig B (2000) Zinc protects against apoptosis of endothelial cells induced by linoleic acid and tumor necrosis factor α. Am J Clin Nutr 71:81–87
Metwalli O, Al-okbi S, Motawi T, El-ahmady O, Abdul-Hafeez S, El-said E (1998) Study of serum metals and lipid profile in patients with acute myocardial infarction. J Islam Acad Sci 11 (1):5–12
Milne DB, Johnson PE (1993) Assessment of copper status: effect of age and gender on reference ranges in healthy adults. Clin Chem 39:883–887
Mocchegianai E, Boemi M, Fumelli P, Fabris N (1989) Zinc-dependent low thymic hormone level in type 1 diabetes. Diabetes 38:932–937
Nadler JL, Malayan S, Loung H, Shaw S, Natarajan RD, Rude RK (1992) Intracellular free magnesium deficiency plays a key role in increased platelet reactivity in type II diabetes mellitus. Diabetes Care 15(7):835–841
Neve J (1991) Methods in determination of selenium states. J Trace Elem Electr Health Dis 5:1–17
Neve J (1996) Selenium as a risk factor for cardiovascular diseases. J Cardiovasc Risk 3:42–47
Noto R, Alicata R, Sfoglian L, Neri S, Bifarella M (1983) A study of cupermia in a group of elderly diabetics. Acta Diabetol Lat 20:81–85
Pham PC, Pham PM, Pham PA, Pham SV, Pham HV, Miller JM, Yanagawa N, Pham PT (2005) Lower serum magnesium levels are associated with more rapid decline of renal function in patients with diabetes mellitus type 2. Clin Nephrol 63:429–436
Rayman MP (2000) The importance of selenium to human health. Lancet 356:233–241
Rayman MP (2005) Selenium in cancer prevention: a review of the evidence and mechanism of action. Proc Nutr Soc 64:527–542
Resnick LM, Altura BT, Gupta RK (1993) Intracellular and extra cellular magnesium depletion in type II (non insulin-dependent) diabetes mellitus. Diabetologia 36:767–770
Rotruck JT, Pope AL, Ganther HE, Swanson AB, Hafeman DG, Hoekstra WG (1973) Selenium: biochemical role as component of glutathione peroxidase. Science 179:538–544
Rude RK (1998) Magnesium deficiency: a cause of heterogeneous disease in humans. J Bone Miner Res 13:749–758
Saari JT (2000) Copper deficiency and cardiovascular disease: role of per oxidation, glycation, and nitration. Can J Physiol Pharmacol 78(10):848–855
Salonen JT, Salonen R, Korpela H, Suntioinen S, Tuomilehto J (1991) Serum copper and the risk of acute myocardial infarction: a prospective population study in men in eastern Finland. Am J Epidemiol 134:268–276
Saris NE, Mervaala E, Karppanen H, Khawaja JA, Lewenstam A (2000) Magnesium: an update on physiological, clinical, and analytical aspects. Clin Chim Acta 294:1–26
Schnack CH, Bauer I, Pregnant P (1992) Hypomagnesaemia in type II (non-insulin dependent) diabetes mellitus is not corrected by improvement of long-term metabolic control. Diabetologia 35:77–79
Shukla N, Maher J, Masters J, Angelini GD, Jeremy JY (2006) Does oxidative stress change ceruloplasmin from a protective to a vasculopathic factor. Atherosclerosis 187(2):238–250
Singh MM, Singh R, Khare A, Gupta MC, Patney NL, Jain VK, Goyal SP, Prakash V, Pandey DN (1985) Serum copper in myocardial infarction-diagnostic and prognostic significance. Angiology 36:504–510
Singh RB, Niaz MA, Rastogi SS, Bajaj S, Gaoli Z, Shoumin Z (1998) Current zinc intake and risk of diabetes and coronary artery disease and factors associated with insulin resistance in rural and urban populations of North India. J Am Coll Nutr 17(6):564–570
Stühlinger HG (2002) Focus on magnesium: magnesium deficiency and cardiovascular disease. Magnesium in cardiovascular disease. J Clin Basic Cardiol 5:55
Suter PM (1999) The effects of potassium, magnesium, calcium and fiber on risk of stroke. Nutr Rev 57(30):84–88

Swanson CA, Longnecker MP, Veillon C, Howe M, Levander OA, Taylor PR (1990) Selenium intake, age, gender, and smoking in relation to indices of selenium status of adults residing in a seleniferous area. Am J Clin Nutr 52:858–862

Tawfeeq FR, Abbas MR, Abdul Kareem Y (2008) Serum copper, zinc and Cu/Zn ratio in diabetics. Iraqi J Commun Med 21(1):64–68

Tiber AM, Sakhaii M, Joffe CD, Ratnaparkhi MV (1986) Relative values of plasma copper, zinc, lipids and lipoproteins as markers for coronary artery disease. Atherosclerosis 62:105–110

Uza G, Pavel O, Kovacis A (1984) Serum concentration of Na, K, Ca, Mg, P, Zn and 2u in patients with essential arterial hypertension. Clin Exp Hypertens A 6(8):1415–1429

Vilanova A, Gutiérrez C, Serrat N, Raga X, Paternain JL (1997) Metallothionein, zinc and copper levels: relationship with acute myocardial infarction. Clin Biochem 30(3):235–238

Walter RM, Uriu-Hare JY, Olin KL (1991) Copper, zinc, magnesium status and complications of diabetes mellitus. Diabetes Care 14:1050–1056

Wang WC, Makela AL, Nanto V, Makela P (1995) Serum selenium levels in diabetic children. A follow-up study during selenium-enriched agricultural fertilization in Finland. Biol Trace Elem Res 47:355–364

Zargar AH, Shah NA, Massodi SR (1998) Copper, zinc and magnesium levels in non insulin dependent diabetes mellitus. Postgard Med J 74:665–668

Chapter 4
Literature Review: Cardiovascular Disorders and Lipid Profile

Gururajan et al. (2010) determined that elevated lipid profile and reduced antioxidants accelerate the formation of atherosclerosis. Multiple lines of evidences have suggested that increased lipids and low antioxidants are the major risk factors for the incidence of acute coronary syndrome. Oxidative stress evaluation is now considered as an index for the assessment of development of CAD. Total cholesterol, LDL cholesterol, HDL cholesterol, and triglycerides were analyzed along with nonenzymatic antioxidants such as vitamin C, vitamin E, reduced glutathione, MDA, and protein thiol in controls and patients with ACS. The levels of total cholesterol and LDL cholesterol were significantly raised in patients when compared to controls in contrast to lowering of HDL-cholesterol levels in patients than controls. Vitamin C, vitamin E, reduced glutathione, MDA, and protein thiol levels were significantly lowered in patients than controls ($P < 0.05$).Oxidative stress and lipid profile should be included as important markers in the early detection of acute coronary syndrome.

Ganesan Karthikeyan et al. (2009) examined the relation between various lipid parameters and occurrence of a first acute MI in Asian men and women. They compared the serum lipid profile in Asian, as compared to non-Asians, with and without acute MI. Serum lipid levels as well as other coronary risk factors were assessed in patients with a first AMI and in control subjects without AMI. The average serum LDL-cholesterol levels were higher in cases of initial AMI from Asia as compared with their corresponding controls. Levels of HDL cholesterol and serum TG were lower in Asian cases of AMI as compared to controls. These patterns were observed in study subjects who were enrolled from South Asia, China/Hong Kong, as well as Southeast Asia. In addition, the average LDL-cholesterol and serum TG levels in both cases and controls from Asia were markedly lower in comparison with non-Asians from the other participating INTERHEART sites. The proportion of cases and controls from Asia who had LDL-cholesterol levels $<$100 mg/dL was 26 % and 32 %, respectively. Differences in serum lipid and lipoprotein patterns between cases of AMI and controls were observed in men and in women and in all age groups examined. While there were differences in the absolute values of serum lipid and serum lipoprotein parameters in the different Asian

N.B. Rizvi and S.A. Nagra, *Minerals and Lipids Profiles in Cardiovascular Disorders in South Asia*, DOI 10.1007/978-3-642-34249-3_4,

populations included in this multinational investigation, the higher serum levels of LDL cholesterol and corresponding lower levels of HDL cholesterol were observed in cases as compared to controls from each of the Asian regions studied. Whereas the estimated risk of AMI was in the expected direction with regard to serum levels of LDL cholesterol (higher risk of acute MI with increasing levels of LDL cholesterol) and HDL cholesterol (lower risk with increasing levels of HDL cholesterol). The results of this large observational study suggest heterogeneity in the serum lipid profile of various segments of the Asian population and that elevated serum levels of LDL cholesterol and lower levels of HDL cholesterol were associated with an increased risk of AMI in Asians. Consistent with the results of prior epidemiological studies, Asian populations typically have a more favorable serum lipid profile than non-Asian populations. Despite their more favorable serum lipid and lipoprotein profile, previously observed associations between various lipid patterns and the risk of AMI were noted in this Asian population. Inasmuch, while the thresholds for the initiation of treatment and "control" of serum lipids and lipoproteins may be considerably lower in Asians than in Caucasians, Asian patients may also acquire benefits from lowering their serum LDL-cholesterol levels and increasing their levels of HDL cholesterol. Similar to other populations throughout the world, attention to the maintenance of optimal weight, through both healthy eating practices and regular physical activity, and use of medications, when necessary, should also pay considerable dividends in Asian as well as in non-Asian populations with regard to lowering their risk of CVD.

Ugwu et al. (2009) compared the lipid profile of diabetic patients and healthy controls. The lipid profiles and lipoprotein levels of known diabetic patients and healthy subjects were studied. Total cholesterol (TC), triacylglycerols (TG), low-density lipoprotein-cholesterol (LDL-C), and high-density lipoprotein-cholesterol (HDL-C) levels were assayed for each group using standard biochemical methods. The mean TC, TG, and low-density lipoprotein-cholesterol levels were lower in the diabetics than in the control subjects though these were not significant ($P > 0.05$). The frequency of high TC level was higher in the diabetic group, while the frequency of low HDL-C level was higher in the healthy controls. The prevalence of high TG and LDL-C was approximately equal in the two groups. The mean ($\pm$SD) HDL-C was significantly lower ($P < 0.05$) in males compared to the females for both diabetic and control groups. The better lipid profiles in the diabetic patients compared to the controls were apparently due to the regime of management of their condition

Kumar and Sivakanesan (2009) observed that the major cause of atherosclerosis, dyslipidemia, acts synergistically with non-lipid-risk factors resulting increase in atherogenesis. Increased (TG) and decreased high-density lipoprotein (HDL cholesterol) and the increased TG/HDL-C ratio are considered as major risk factors in the development of insulin resistance and metabolic syndrome. The accuracy of TG/HDL-cholesterol ratio in predicting coronary heart disease (CHD) risk is not properly established by recent research. The study was undertaken to evaluate the usefulness of lipid ratios of TC/HDL cholesterol, TG/HDL cholesterol, and LDL/HDL cholesterol in predicting CHD risk in normo-lipidemic patients with myocardial infarction and to compare the results with healthy subjects. Total cholesterol, TC/HDL-C ratio, triglycerides, LDL cholesterol, and LDL-cholesterol/HDL-cholesterol ratio were higher in MI patients ($P < 0.001$). HDL-C concentration was significantly lower in MI patients than controls ($P < 0.001$). Higher ratio of TC/HDL cholesterol, TG/HDL cholesterol, and LDL cholesterol/HDL cholesterol was observed in AMI patients compared to controls.

Karthikeyan et al. (2009) determined the prevalence of lipid and lipoprotein abnormalities and their association with the risk of a first acute myocardial infarction (AMI) among Asians. Patterns of lipid abnormalities among Asians and their relative impact on cardiovascular risk have not been well characterized. Among both cases and controls, mean low-density lipoprotein-cholesterol (LDL-C) levels were about 10 mg/dL lower in Asians compared with non-Asians. High-density lipoprotein-cholesterol (HDL-C) levels were slightly lower among Asians compared with non-Asians. There was preponderance of people with low HDL-C among South Asians (South Asia vs. rest of Asia: cases 82.3 % vs. 57.4 %; controls 81 % vs. 51.6 %; $P < 0.0001$ for both comparisons). However, despite these differences in absolute levels, the risk of AMI associated with increases in LDL-C and decreases in HDL-C was similar for Asians and non-Asians. Among South Asians, changes in apolipoprotein (Apo)A1 predicted risk better than HDL-C. ApoB/ApoA1 showed the strongest association with the risk of AMI. The preserved association of LDL-C with risk of AMI among Asians, despite the lower baseline levels, suggested the need to rethink treatment thresholds and targets in this population. The low HDL-C level among South Asians requires further study and targeted intervention.

Idemudia and Ugwuja (2009) reported that the association between hypertension and dyslipidemia is well established and both may add up to increase patients' susceptibility to the development of coronary heart disease. Of the 150 hypertensive patients, 54 % ($n = 69$) were females with majority (45.7 %) in the age range 50–59 years, while majority (53.6 %) of hypertensive males were in the age group 40–49 years. Hypertensive patients have significantly higher lipid profile except for HDL cholesterol, which did not show any significant difference in the two groups. Among the hypertensive patients, total cholesterol was positively correlated with triglyceride (0.399, $P < 0.05$), LDL cholesterol ($r = 0.609$, $P < 0.05$), and HDL cholesterol ($r = 0.866$, $P < 0.05$), and HDL-C positively correlated with LDL-C ($r = 0.218$, $P < 0.05$). In normotensive patients, however, LDL cholesterol was negatively correlated with triglyceride ($r = -0.409$, $P < 0.05$), while total cholesterol was positively correlated with LDL cholesterol ($r = 0.876, P < 0.05$). Hypertensive Nigerians have significantly elevated plasma total cholesterol, triglyceride, and LDL-C but comparable HDL-C with normotensives. The clinical implications of elevated HDL-C in hypercholesterolemic hypertensive Nigerians are unclear.

Smith and Lall (2008) reported that in India, diabetes is not an epidemic anymore but has turned into a pandemic, according to the International Journal of Diabetes in developing countries which labeled India the diabetes capital of the world. Mainly because India now has the highest number of diabetic patients in the world, the International Diabetes Federation estimates that the number of diabetic patients in India more than doubled from 19 million in 1995 to 40.9 million in 2007. It is projected to increase to 69.9 million by 2025. Currently, up to 11 % of India's urban population and 3 % of rural population above the age of 15 have diabetes. The most prevalent is type II diabetes, which constitutes 95 % of the diabetic population in the country. Indian population faces higher risk for diabetes and its complications. The result revealed that serum total cholesterol, LDL-C, and triglycerides were significantly raised ($P < 0.0001$) whereas the level of HDL-C was significantly lower ($P < 0.0001$) in diabetic subjects as compared to control. There is a significant change found between diabetic male and female serum lipid profile. Diabetic male has highly significant higher level of cholesterol, LDL-C, and triglycerides and significantly lower level of HDL-C in the diabetic female. It was concluded that hypercholesterolemia, hypertriglyceridemia, and lipoprotein are the main lipid abnormalities found in diabetes which is risk for coronary artery disease. In diabetes, sex plays a significant effect on risk of coronary artery disease.

Mosca et al. (2004) reported that worldwide, cardiovascular disease is the largest single cause of death among women, accounting for one third of all deaths. In many countries, including the United States, more women than men die every year of CVD, a fact largely unknown by physicians. The public health impact of CVD in women is not related solely to the mortality rate, given that advances in science and medicine allow many women to survive heart disease. For example, in the United States, 38.2 million women (34 %) are living with CVD, and the population at risk is even larger. In China, a country with a population of approximately 1.3 billion, the age-standardized prevalence rates of dyslipidemia and hypertension in women 35–74 years of age are 53 % and 25 %, respectively, which underscores the enormity of CVD as a global health issue and the need for prevention of risk factors in the first place. As life expectancy continues to increase and economies become more industrialized, the burden of CVD on women and the global economy will continue to increase. The 2004 guidelines emphasized the importance of recognizing the spectrum of CVD and thus classified women as being at high risk, intermediate risk, lower risk, and optimal risk. Classification was based on clinical criteria and/or the Framingham global risk score. The expert panel suggested several gaps in knowledge related to the prevention of CVD that must be addressed to optimize the cardiovascular health of women. More rigorous testing of the impact of guidelines themselves on preventing risk factors, slowing the progression of risk factors, and reducing the burden of CVD is needed. The development and testing of effective methods to implement guidelines in various health-care settings, at work sites, and in communities are also research priorities. The role of communication of risk and barriers to CVD prevention should be studied and incorporated into creative methods to disseminate and implement guidelines among diverse populations of women.

Mengesha (2006) determined the serum lipid profile of diabetes mellitus (DM) patients receiving treatment at Gaborone City Council clinics. A total of 401 patients were studied over a 3-month period. It was found that 33.5 % had hypercholesterolemia and 38.9 % hypertriglyceridemia. The mean low-density lipoprotein (LDL-C) levels were higher in females than in males, but there was no difference in LDL-C levels between type I and II DM patients. There was no difference in cholesterol, triglyceride, and high-density-lipoprotein (HDL-C) levels between genders or between type I and II patients. Hyperlipidemia was associated with high body mass index. Only hypertriglyceridemia was associated with high blood pressure. Hyperlipidemia was not associated with exercise, smoking, or alcohol consumption in the DM patients studied.

Saha et al. (2006) conducted a prospective study in the northern region of Bangladesh to investigate the serum lipid profile, namely, the level of total cholesterol (TC), triglyceride (TG), HDL cholesterol, and LDL cholesterol of hypertensive patients, and compared them with levels of control subjects. The results revealed that serum total cholesterol, triglyceride, and LDL cholesterol were significantly markedly raised ($P > 0.001$), whereas the level of HDL cholesterol was significantly lower ($P > 0.001$) in hypertensive patients as compared to control subjects. No significant changes of serum lipid profile were found between male and female hypertensive patients, but in control subjects, markedly higher levels of serum lipid profile was observed in male compared to that of female. It was concluded that hypercholesterolemia, hypertriglyceridemia, and low-density lipoprotein are the main lipid abnormalities on the incidence of hypertension in the study area.

Shahid Habib (2006) aimed to study the prevalence of desirable and high-risk levels of lipid profiles in Saudi type II diabetics according to ATP III guidelines. Fasting blood samples were analyzed for total cholesterol (TC), triglycerides (TG), low-density lipoprotein cholesterol (LDL-C), high-density lipoprotein cholesterol (HDL-C), and glucose and glycosylated hemoglobin (HbA1c). They assessed the percentage of patients falling into desirable, borderline, and high-risk categories according to the criteria laid down by Adult Treatment Panel III of American Medical Association. It was found that 56.6, 23.6, 77.1, and 48.9 % of diabetic subjects had borderline to high-risk levels of TC, TG, LDL cholesterol, and HDL cholesterol, respectively. It was concluded that type II diabetic patients have a high frequency of atherogenic dyslipidemia especially for TC, LDL-C, and HDL-C. It is suggested that along with glycemic control, physicians should focus more on lipid profiles also.

Mackay and Mensah (2004) reported that heart disease and stroke are currently the leading causes of death in all developed countries and in most developing countries. Heart disease and stroke are no longer diseases of old men in developed countries. They are also diseases of women, young adults, and even children. Cardiovascular diseases are more than just health problems. Yet worldwide, most people who have risk factors are either not treated or are inadequately treated. Special attention to high blood pressure, high blood cholesterol, tobacco, and other major risk factors is crucial.

Iqbal et al. (2004) investigated the changes in total cholesterol, high-density lipoprotein (HDL) cholesterol, low-density lipoprotein (LDL) cholesterol, and triglycerides in serum of Pakistani patients with AMI due to age, gender, body mass index (BMI), diabetes, hypertension, and smoking and also find out the prevalence of hypercholesterolemia, hypertriglyceridemia, "low HDL cholesterol," and "isolated low HDL cholesterol" in them. LDL cholesterol was determined using the Friedewald formula. Mean serum concentrations of total cholesterol, HDL-C, LDL-C, and triglycerides in AMI patients were found to be 181 $\pm$ 50 mg/dL, 35.7 $\pm$ 11.3 mg/dL, 110 $\pm$ 47 mg/dL, and 177 $\pm$ 127 mg/dL, respectively. Mean levels of total cholesterol and HDL cholesterol were not significantly affected by age, gender, BMI, diabetes mellitus, hypertension, and smoking. Mean LDL-cholesterol concentration, however, was found to be significantly increased in diabetes mellitus ($P = 0.047$), while age, gender, BMI, hypertension, and smoking had no significant effect on the levels of this lipoprotein. Mean levels of triglycerides were significantly decreased in older patients (>50 years) compared to younger (<50 years) ones ($P = 0.01$ g). Gender, BMI, diabetes mellitus, hypertension, and smoking, however, had no effect on triglyceride levels. The frequencies of hypercholesterolemia, hypertriglyceridemia, "low HDL cholesterol," and "isolated low HDL cholesterol" were found to be 30.6 %, 30.1 %, 48.6 %, and 34.1 %, respectively. Conclusion: High prevalence of hypertriglyceridemia and low HDL cholesterol (which constitute a component of metabolic syndrome) in Pakistani AMI patients is suggestive that these two lipid abnormalities could be playing a major role in the development of atherosclerosis in Pakistani population.

Haffner (2004) reported that the most common pattern of dyslipidemia in patients with type II diabetes patients is elevated triglyceride levels and decreased HDL-cholesterol levels. The mean concentration of LDL cholesterol in those with type II diabetes is not significantly different from that in those individuals who do not have diabetes. However, qualitative changes in LDL cholesterol may be present. In particular, patients with diabetes tend to have a higher proportion of smaller and denser LDL particles, which are more susceptible to oxidation and may thereby increase the risk of cardiovascular events. Insufficient data are available to make recommendations on the measurement of particle size in clinical practice. As in those who do not have diabetes, lipid levels may be affected by factors unrelated to glycemia or insulin resistance, such as renal disease, hypothyroidism, and frequent occurrence of genetically determined lipoprotein disorders (e.g., familial combined hyperlipidemia and familial hypertriglyceridemia). These genetic disorders may contribute to the severe hypertriglyceridemia seen in some patients with diabetes. Furthermore, use of alcohol or estrogen may also contribute to hypertriglyceridemia. Aggressive therapy of diabetic dyslipidemia will reduce the risk of CVD in patients with diabetes. Primary therapy should be directed first at lowering LDL levels. The goal is to reduce LDL concentrations to $\leq$100 mg/dL [2.60 mmol/L]. The initiation level for behavioral interventions is also an LDL-C of $\geq$100 mg/dL (2.60 mmol/L). If the HDL is <40 mg/dL, a fibric acid, such as fenofibrate, or niacin might be used in patients with LDL cholesterol between 100 and 129 mg/dL. The initial therapy for hypertriglyceridemia is improved glycemic control and lifestyle intervention. Additional triglyceride lowering can be achieved with fibric acid derivatives (gemfibrozil or fenofibrate) or niacin. For subjects with both high LDL and triglyceride levels, high-dose statin may be used.

Johnson et al. (2004) estimated the prevalence of concurrent hypertension and dyslipidemia among a general veteran population and separately among patients with diabetes mellitus and compared the prevalence of cardiovascular disease among groups with isolated versus concurrent hypertension and dyslipidemia. The proportion of patients with isolated or concurrent hypertension and dyslipidemia was estimated based on diagnostic, pharmacy, laboratory, and vital sign information, and the age-adjusted proportions of individuals with cardiovascular disease were compared between groups.

They found that 57.8 % of all patients had hypertension or dyslipidemia; 30.7 % had both. Sixteen percent of all patients had diabetes mellitus, and 66.3 % of these patients had concomitant hypertension and dyslipidemia. The prevalence of coronary artery disease was often more than doubled among patients with concomitant conditions compared with patients with either condition alone. The prevalence of stroke and peripheral arterial disease similarly increased among patients with both conditions. The prevalence of these cardiovascular diseases was highest among patients with diabetes mellitus. The prevalence of cardiovascular disease was high among this population of older, predominately male US veterans.

Wilsgaard and Arnesen (2004) observed that the steady increase in body weight is becoming a major health problem in Western societies. They assessed the association between 8-year change in body weight and serum lipids in a population-based study comprising of men and women aged 20–61 years at baseline in 1986. Comparisons between different strata of age, sex, initial weight, and categories of smoking status change were also addressed. Significant associations between body mass index (BMI) change and change in high-density lipoprotein (HDL) cholesterol, total cholesterol, and triglycerides were observed in all 10-year age groups both in men and women. The weakest associations were observed in persons older than 50 years of age, and the associations were also weaker in women than in men. In quartile groups of baseline BMI, a significant linear trend was observed for HDL cholesterol in men and for total cholesterol in both men and women. The associations were less adverse for persons in a higher quartile group of baseline BMI. The association between BMI change and serum lipid change was strongest for persons who were consistent smokers or nonsmokers at each survey. They concluded that an increase in BMI has been shown to be associated with adverse changes in serum lipids. The associations were weaker in women than in men.

Ellison et al. (2004) reported while genetic factors are major determinants of high-density lipoprotein cholesterol (HDL-C), environmental factors also play a role. The latter include three modifiable lifestyle factors: alcohol consumption, physical activity, and smoking. They compared the relative effects of alcohol, physical activity, and smoking on HDL-C levels, using data from subjects (1,226 women and 1,083 men), and aged 25–91 years, from randomly selected families participating in the National Heart, Lung, and Blood Institute Family Heart Study. Alcohol consumption was associated with the largest increment in HDL-C (an increase of 9.0–13.1 mg/dL from nondrinker to highest categories); physical activity with a more modest increment (an increase of 3.0–3.3 mg/dL from lowest to highest categories); and cigarette smoking with a large decrement in women (a decrease of 9.9 mg/dL) and a modest one in men (a decrease of 2.6 mg/dL) between nonsmoker and $\geq$20 cigarettes per day categories. The three lifestyle behaviors plus age, body mass index, education, and current estrogen use explained 22.4 % and 18.2 % of the total variance of HDL-C for women and men, respectively. Alcohol accounted for 28.6 % of this variance among women and 50.1 % among men; smoking accounted for 6.7 % and 3.3 %, respectively; and physical activity for 2.7 % and 3.6 %, respectively, among women and men. Age, body mass index, education, and current estrogen use explained the remaining 62.0 % and 43.0 %, respectively, of the variance attributed to environmental factors. This study suggested that, among lifestyle behaviors, alcohol consumption is the more important correlate of HDL cholesterol.

Burman et al. (2004) evaluated the effect of dietary fat on plasma lipoprotein (a) [Lp(a)] levels and studied the potential of Lp(a) as a more reliable marker for CAD compared to other lipids and lipoproteins. Results indicated that plasma Lp (a), total serum cholesterol (TC), triglyceride (TG), low-density lipoprotein cholesterol (LDL-C), high-density lipoprotein cholesterol (HDL-C), and LDL-C/HDL-C ratio of CAD patients were significantly higher than the controls. High fat intake was found to be associated with higher plasma Lp(a) levels ($P < 0.05$) in patients only. No significant correlation was found between Lp(a) levels and other conventional lipoproteins. The lack of correlation between Lp(a) and other lipoproteins indicates its potential as an independent risk factor for CAD. High fat intake led to higher plasma Lp(a) levels in patients; hence, it would be worthwhile to evaluate the effect of quality and quantity of fat intake on plasma Lp(a) levels in a larger sample size.

Nigam et al. (2004) estimated serum lipids and lipoproteins in patients with acute myocardial infarction during acute phase (day 1, 2, and 3), predischarge and after 3 months. Serum total lipids, total cholesterol (TC), and LDL cholesterol (LDL-C) showed no significant change during the hospital stay and 3-month follow-up. HDL-C, however, started falling from day 2 onward with statistically significant reduction at predischarge and remained so at 3 months. The ratios of TC/HDL-C and LDL-C/HDL-C showed significant increase on predischarge day as compared to day 1. Serum triglycerides also showed an increasing trend after myocardial infarction with a significant increase on day 3 and predischarge as compared to day 1. They concluded that the optimum time for assessment of serum lipid profile in patients with myocardial infarction seems to be within 24 h of the acute episode.

Wenger (2003) reported that during the past decade, an overall theme has emerged, validating the exploration of gender-based differences in coronary heart disease (CHD) as a basis for clinical strategies to improve outcomes for women. Underrepresentation of women in most of CHD and lack of gender-specific reporting in many clinical trials continue to limit the available knowledge and evidence-based medicine needed to devise optimal managements for women with CHD. Control of conventional coronary risk factors provides comparable cardioprotection for men and women. Current evidence fails to show cardiac protection from menopausal hormone therapy. Clinical presentations of coronary heart disease (CHD) and management strategies differ between the sexes. Underutilization of proven beneficial therapies is a contributor to less favorable outcomes in women. The contemporary increased application of appropriate diagnostic, therapeutic, and interventional managements has favorably altered the prognosis for women, particularly when the data are adjusted for baseline characteristics. Better education of women during office visits, earlier and more aggressive control of coronary risk factors, and a greater index of suspicion regarding chest pain and its appropriate evaluation may help to reverse the trend of late referral and late intervention. Research indicates that behavioral changes on the part of women and reshaping of practice patterns by their health-care providers may dramatically reduce the number of women disabled and killed by CHD each year.

According to AHA (2003), one of the enduring half-truths about CVD is that heart disease is man's disease. The fact is cardiovascular diseases are devastating to women too. In terms of total deaths, in every year since 1984, CVD has claimed the lives of more females than males. The harsh fact is cardiovascular diseases are the no.1 killer of women and men. Starting at age 75, the prevalence of CVD among women is higher than among men. Another myth is that CVD only affects old people, while other people are at higher risk.

Oparil et al. (2003) reported that a clearer understanding of the pathogenesis of hypertension will probably lead to more highly targeted therapies and to greater reduction in hypertension-related cardiovascular disease morbidity than can be achieved with current empirical treatment. Endothelial dysfunction, increased vascular reactivity, and vascular remodeling may be causes, rather than consequences, of blood pressure elevation; increased vascular stiffness contributes to isolated systolic hypertension in the elderly. Essential hypertension, or hypertension of unknown cause, accounts for more than 90 % of cases of hypertension. The complexity of pathophysiological mechanisms that lead to blood pressure elevation is such that selective, mechanistically based antihypertensive treatment is rarely possible in any hypertensive patient. Hypertension is highly prevalent among middle-aged and elderly persons in our population, and the success rate in controlling blood pressure in these individuals is poor. Current treatment guidelines generally recommend a generic approach to treating hypertension, with little emphasis on selecting therapy on the basis of the underlying pathophysiology of the elevated blood pressure. With increased recognition of specific causes, it may be possible to develop therapies selective for distinct pathophysiological mechanisms with fewer adverse effects, resulting in more effective blood pressure reduction.

Seyoum et al. (2003) determined serum lipids and lipoproteins in randomly selected diabetic patients attending the Tikur Anbessa Hospital diabetic clinic. The main objective of the study was to analyze lipid levels in type I and type II diabetic patients. 53 % were males and 47 % were females. There was 46.4 % type I and 53.6 % type II patients. Hypercholesterolemia and hypertriglyceridemia were seen in 18.5 % and 14.2 % of the patients. Total cholesterol was significantly higher in females than in males and in type II than in type I patients (179.3 ± 48.4 mg/dL vs. 154.1 ± 38.2 mg/dL, $P < 0.01$ and 183.2 ± 43.7 mg/dL vs. 145.9 ± 37.6 mg/dL, $P < 0.001$, respectively). Triglycerides and LDL cholesterol were also significantly higher in type II diabetic patients than in type I diabetic patients (162.7 ± 10.5 mg/dL vs. 91.5 ± 53.3 mg/dL, $P < 0.001$ and 105.6 ± 36.2 mg/

dL vs. 81.9 ± 32.2 mg/dL, $P < 0.001$), but HDL cholesterol was the same in both types of diabetic patients. Similarly, hyperlipidemia was associated with obesity and hypertension. The study confirms that lipid values are high particularly in type II diabetic patients. Hence, our patients are at increased risk of developing atherosclerosis; therefore, periodic checkup of lipids in diabetic patients and effective treatment of the dyslipidemia along with a tight metabolic control were recommended.

Tsimikas et al. (2003) conducted the study to test the hypothesis that plasma markers of oxidized low-density lipoprotein (OxLDL) reflect acute coronary syndromes (ACS). Oxidized LDL contributes to the pathogenesis of atherosclerosis, but its role in ACS is not established. Baseline OxLDL IgG autoantibody levels were higher in the MI group ($P < 0.0001$). At 30-day follow-up, the mean IgM OxLDL titers increased by 48 % ($P < 0.001$) and 20 % ($P < 0.001$), and IgM LDL-IC increased by 60 % ($P < 0.01$) and 26 % ($P < 0.01$) in the MI and UA groups, respectively. The OxLDL-E06 levels increased by 54 % ($P < 0.01$) in the MI group at hospital discharge and by 36 % at 30 days. No significant changes in any OxLDL markers were noted in the other groups. The OxLDL-E06 levels strongly paralleled the acute rise in lipoprotein(a), or Lp(a), in the MI group, suggesting that toxic OxPL are preferentially bound to Lp(a). Oxidized LDL-E06 also correlated with Lp(a). Circulating OxLDL-specific markers strongly reflect the presence of ACS, implying immune awareness to newly exposed oxidation-specific epitopes and possible release of OxLDL in the circulation. The OxLDL-E06 measurements provide novel insights into plaque rupture and the potential atherogenicity of Lp(a).

Bermudez et al. (2002) assessed the disparity in prevalence of type II diabetes among population groups underscored dietary and plasma risk factors for cardiovascular disease. Plasma lipids and apolipoproteins and dietary intakes of macronutrients were measured in elderly subjects (60–98 years). Differences in plasma lipids due to diabetes were assessed among the Hispanics. Intakes of carbohydrate and polyunsaturated fatty acids were higher, and intakes of cholesterol and saturated and monounsaturated fatty acids were lower in Hispanics than in non-Hispanic whites. Concentrations of total cholesterol, HDL-C, and apolipoprotein A-I were significantly lower among Hispanic women than among non-Hispanic white women; a similar trend was seen in men. Dyslipidemia (high triacylglycerols and low HDL cholesterol) was more prevalent among Hispanics with than without diabetes. Ethnic differences in serum lipids exist and appear to be associated with differences in dietary intakes. However, both Hispanics and non-Hispanic whites had lipid profiles indicating a high risk of cardiovascular disease. Hispanics with diabetes were at higher risk of dyslipidemia than were those without diabetes. They suggested that lifestyle changes, including diet modification and exercise, could be of significant benefit to both ethnic groups

Saaddine et al. (2002) reported that 18.0 % of participants (95 % CI, 15.7–22.3 %) had poor glycemic control (hemoglobin A_{1c} level > 9.5 %), and 65.7 % (CI, 62.0–69.4 %) had blood pressure less than 140/90 mmHg. Cholesterol was monitored biannually in 85.3 % (CI, 83.1–88.6 %) of participants, but only

42.0 % (CI, 34.9–49.1 %) had LDL-cholesterol levels less than 3.4 mmol/L (<130 mg/dL). Persons taking insulin were more likely than those who were not to have annual dilated eye examination (72.2 % [CI, 66.3–78.1 %] vs. 57.6 % [CI, 53.7–61.5 %]) and foot examination (67.3 % [CI, 61.4–73.2 %] vs. 47.1 % [CI, 43.2–51.0 %]) but were also more likely to have poor glycemic control (24.2 % [CI, 18.3–30.1 %] vs. 15.5 % [CI, 11.6–19.4 %]). According to US data collected during 1988–1995, a gap exists between recommended diabetes care and the care patients actually receive. These data offer a benchmark for monitoring changes in diabetes care.

Nthangeni et al. (2002) determined the dietary intake, practices, knowledge, and barriers to dietary compliance of black South African type II diabetic patients attending primary health-care services in urban and rural areas. Reported dietary results indicated that mean energy intakes were low (<70 % of Recommended Dietary Allowance), 8,086–8,450 kJ/day and 6,967–7,382 kJ/day in men and women, respectively. Urban subjects had higher ($P < 0.05$) intakes of animal protein and lower ratios of polyunsaturated fat to saturated fat than rural subjects. The energy distribution of macronutrients was in line with the recommendations for a prudent diet, with fat intake less than 30 %, saturated fat less than 10 %, and carbohydrate intake greater than 55 % of total energy intake. In most respects, nutrient intakes resembled a traditional African diet, although fiber intake was low in terms of the recommended 3–6 g/1,000 kJ. Poor glycemic control was found in both urban and rural participants, with more than half of subjects having fasting plasma glucose above 8 mmol/L and more than 35 % having plasma glycosylated hemoglobin level above 8.6 %. High triglyceride levels were found in 24–25 % of men and in 17–18 % of women. Obesity (body mass index $\geq$ 30 kg/m^2) was prevalent in 15–16 % of men compared with 35–47 % of women; elevated blood pressure ($\geq$160/95 mmHg) was least prevalent in rural women (25.9 %) and most prevalent in urban men (42.4 %). The majority of black, type II diabetic patients studied showed poor glycemic control. Additionally, many had dyslipidemia, were obese, and/or had an elevated blood pressure. Quantitative and qualitative findings indicated that these patients frequently received incorrect and inappropriate dietary advice from health educators.

Youmbissi et al. (2001) studied hypertensive black African Cameroonians and normotensive controls, age and sex matched. In groups, total cholesterol, HDL cholesterol, LDL cholesterol, triglycerides, and apolipoproteins A-l (apo A-I) and B (apo B) were assayed in the peripheral blood. All lipids fractions were assayed using enzymatic methods except for apolipoproteins which were measured by immunoturbidimetry. Fisher test and correlation coefficient were used in the statistical analysis. Hypertensive subjects showed significantly higher levels of total cholesterol, HDL cholesterol, LDL cholesterol, and triglycerides and lower levels of apolipoproteins. In hypertensive women, triglycerides levels were significantly correlated to the severity of the hypertension ($P < 0.05$). Those hypertensive men, who were being treated with diuretics, beta-blockers, and calcium antagonists, had higher lipids levels, and these were correlated to the length of the treatment. There was no correlation between apo A-I and HDL levels or between apo B and LDL levels. Most dyslipidemia were hypercholesterolemia (53 %) followed by hypertriglyceridemia (2 %) and mixed hyperlipidemias (23 %).

Kanaley et al. (2001) reported age-related increases in total body fat, but the impact of menopause on abdominal fat distribution is still unclear. The purpose of the study was to determine the impact of menopausal status on abdominal fat distribution using magnetic resonance imaging (MRI). In addition, they also investigated the influence of abdominal fat distribution on blood lipid profiles and leptin concentrations. The percent visceral abdominal fat was significantly lower ($P < 0.05$) in the premenopausal women than in either postmenopausal group (PRE, 23.2 % $\pm$ 1.7 %; POST, 28.9 % $\pm$ 1.8 %; ERT, 28.9 % $\pm$ 1.6 %). Menopausal status and age did not influence any of the blood lipid values. Abdominal fat distribution was a significant predictor of cholesterol concentrations and the cholesterol/high-density lipoprotein-cholesterol (HDL-C) ratio, but only accounted for approximately 15 % of the variability in these levels. Total body fat and physical activity accounted for 47 % of the variability in leptin concentrations, while abdominal fat distribution, age, and menopausal status were not significant predictors. In conclusion, in early postmenopausal women, the level of physical activity accounts for the variability in abdominal fat distribution observed, while menopausal status and age do not play a significant role. ERT was not associated with additional benefits in abdominal fat distribution compared with postmenopausal women not on ERT or in the blood lipid profile in these women.

Njelekela et al. (2001) examined the prevalence of selected risk factors according to the World Health Organization (WHO) CARDIAC Study protocol and compared them with a similar study conducted more than a decade ago. The measured weight, body mass index (BMI), and prevalence of obesity (BMI $\geq$ 30 kg/m^2) increased significantly among women in the 1998 survey in rural Handeni and urban Dar. The overall prevalence of obesity was higher for women in the most recent survey (22.8 %, $P < 0.0001$). Diastolic blood pressure (DBP) was higher in the most recent survey for women in Handeni. The overall prevalence of hypertension (blood pressure > 160/95 mmHg, or antihypertensive drug use) rose to 41.1 % in 1998 ($P < 0.001$) for men and to 38.7 % ($P < 0.05$) for women. The mean total serum cholesterol and prevalence of hypercholesterolemia increased significantly in the most recent survey in the three studied areas. The overall prevalence of hypercholesterolemia (serum cholesterol > 5.2 mmol/L) was higher in the 1998 survey for both men (21.8 %, $P < 0.0001$) and women (54.0 %, $P < 0.0001$). The mean HDL cholesterol increased significantly in the most recent survey, with a significant reduction in the mean atherogenic index, though these were still at higher levels (men 5.8, $P < 0.0001$; women 5.1, $P < 0.0001$ vs. 1987). A strong positive correlation was observed between blood pressure (SBP and DBP) and body mass index, total serum cholesterol, and sodium-to-potassium ratio. These data suggested that for the past decade, there has been an increase in the mean levels and prevalence of selected cardiovascular risk factors in Tanzania.

Gaziano et al. (1999) sought to evaluate the potential interactions between systemic hypertension and blood lipids on the risk of myocardial infarction (MI).

Recent evidence suggested that hypertension may interact with other risk factors such as dyslipidemia in the development of CAD. However, the precise nature of that interrelation remains unclear. The age- and sex-adjusted OR of MI was 1.61 (95 % confidence interval [CI] 1.15–2.25) among treated hypertensive compared with non-hypertensive. Further adjustment for coronary risk factors did not materially alter the results (OR 1.67, 95 % CI 1.16–2.41). The apparent risk associated with hypertension was substantially attenuated by the addition of either high-density lipoprotein cholesterol (OR 1.25, 95 % CI 0.82–1.90) or triglycerides (OR 1.37, 95 % CI 0.91–2.05). No significant interactions were found between hypertension and any lipoprotein parameter. These data indicate that the risk of MI associated with treated hypertension may have a lipid mechanism involving high-density lipoprotein cholesterol and/or triglycerides.

Gould et al. (1998) determined the effect of incorporating the results of eight recently published trials of Hmg-CoA reductase inhibitors ("statin") on the conclusions from the previously published meta-analysis regarding the clinical benefit of cholesterol lowering. They used the same analytical approach as in previous investigation, separating the specific effects of cholesterol lowering from the effects attributable to the different types of intervention studied. Including the statin trial findings into the calculations led to a prediction that for every 10 percentage points of cholesterol lowering, CHD mortality risk would be reduced by 15 % ($P < 0.001$), and total mortality risk would be reduced by 11 % ($P < .001$), as opposed to the values of 13 % and 10 %, respectively. Cholesterol lowering in general and by the statin in particular does not increase non-CHD mortality risk. The relationships (slope) between cholesterol lowering and reduction in CHD and total mortality risk became stronger, and the standard error of the estimated slopes decreased by about half. Use of statin does not increase non-CHD mortality risk. The effect of the statin on CHD and total mortality risk can be explained by their lipid-lowering ability and appears to be directly proportional to the degree to which they lower lipids.

Ballantyne (1998) reported that based on the established relation between low-density lipoprotein (LDL) cholesterol and coronary artery disease (CAD), the treatment guidelines of the US National Cholesterol Education Program (NCEP) focus on LDL-cholesterol reduction for primary and secondary prevention of CAD events. Abundant clinical trial evidence supports the importance of LDL-cholesterol lowering in decreasing CAD risk, both in angiographic trials, which measure CAD progression, and in trials with morbidity and mortality endpoints. Even in trials of statin therapy, in which substantial reductions of LDL cholesterol have been obtained, statin decrease (by 23–37 %) but do not entirely eliminate events, suggesting that lipid parameters besides LDL cholesterol, such as high-density lipoprotein (HDL) cholesterol, triglyceride, lipoprotein(a), and LDL particle size and susceptibility to oxidation, as well as other risk factors, influence CAD risk. Systematic screening to identify high-risk patients and methodical follow-up to implement diet, lifestyle modification, and drug therapy to lower LDL cholesterol, as provided for in the NCEP guidelines, should lead to significant benefits in the prevention of CAD events.

Mgonda et al. (1998) assessed insulin sensitivity using a glucose–insulin infusion test in newly diagnosed nonobese hypertensive black Tanzanians with normal glucose tolerance and in normotensive control subjects matched for age, sex, and body mass index. The steady-state blood glucose and metabolic clearance rate of glucose (MCR) were used as measures of insulin sensitivity. The mean MCR (glucose) was significantly reduced (7.12 ± 0.57 vs. 9.50 ± 0.69 μmol/kg/min; $P < 0.05$), and mean steady-state blood glucose was significantly elevated (5.0 ± 0.3 vs. 3.7 ± 0.3 mmol/L; $P < 0.01$) in subjects with hypertension compared with the normotensive group. For all subjects, there was a significant inverse correlation between MCR (glucose) and systolic ($P = 0.003$) and diastolic ($P = 0.005$) blood pressure, and a positive correlation was found between fasting serum insulin levels and systolic ($P = 0.005$) and diastolic ($P = 0.004$) blood pressure. These observations were independent of body mass index and serum lipid levels. These data indicated a strong association between insulin-mediated glucose uptake and blood pressure in this population of normal weight-untreated urban Africans.

Singh et al. (1997) determined the age-specific prevalence of hypertension and blood pressure (BP) levels in relation to diet and lifestyle factors in North Indians. Diagnosis of hypertension was based on new World Health Organization/International Society of Hypertension (WHO/ISH) criteria. Risk factors were assessed based on WHO guidelines. The prevalence of hypertension according to WHO/ISH criteria was 23.7 % and by old WHO criteria 13.3 %. In the WHO/ISH hypertensive group, isolated diastolic hypertension was present in 47.3 % males and 40.6 % females. Males have a slightly higher prevalence than females in the young age group; however, the prevalence rates are comparable in the older age groups. In both sexes, the prevalence rates and BP level increased with older age. Multivariate analysis revealed that age, higher body mass index, central obesity, and higher socioeconomic status were independently and strongly associated with hypertension in both sexes. Higher dietary fat and salt intake and lower physical activity were weakly but significantly associated with hypertension. Association of higher socioeconomic status, higher body mass index, and central obesity in North Indian adults with higher fat intake, lower physical activity, and higher prevalence and level of hypertension indicated that these populations may benefit by decreasing the dietary fat intake and increasing physical activity, with an aim to decrease central obesity for decreasing hypertension in North Indians.

Sixth Reports of the Joint National Committee (1997) provided guidance for primary care clinicians. The committee recognized that the responsible clinician's judgment of the individual patient's needs remains paramount. Therefore, this national guideline should serve as a tool to be adapted and implemented in local and individual situations. Using evidence-based medicine and consensus, the report updates contemporary approaches to hypertension control. Among the issues covered are the important need for prevention of high blood pressure by improving lifestyles, the cost of health care, the use of self-measurement of blood pressure, the role of managed care in the treatment of high blood pressure, the introduction of new combination antihypertensive medications and angiotensin II receptor blockers, and strategies for improving adherence to treatment. The JNC VI report places more emphasis than earlier reports on absolute.

Khoo et al. (1997) reported that over a thousand subjects visiting a city private medical clinic for health screening were advised for their lipid profile and other cardiovascular risk factors. The prevalence of hyperlipidemia was moderately high. Of the subjects studied, 58.5 % had elevated serum cholesterol, 14.8 % had raised triglycerides, 64.9 % had raised LDL cholesterol, and 20.8 % had low HDL cholesterol. Male subjects showed higher mean values and abnormality frequency in TC, TG, LDL cholesterol, and TC/HDL cholesterol and LDL/HDL cholesterol as compared to female subjects. Although significant ethnic differences were not detected for certain lipid parameters (e.g., TC, TG, and HDL cholesterol), the Indians appeared to have higher mean lipid values (except HDL cholesterol) and higher percentage abnormality for all the lipid parameters as compared to the Chinese and the Malays. In correlation studies, the following lipid parameters were positively correlated: TC versus TG, LDL cholesterol, and TC/HDL cholesterol; TG versus TC/HDL cholesterol and LDL/HDL cholesterol; and LDL cholesterol versus TC/HDL cholesterol and LDL/HDL cholesterol. On the other hand, TC versus HDL cholesterol; TG versus HDL cholesterol, LDL cholesterol, and HDL cholesterol; and HDL cholesterol versus TC/HDL cholesterol and LDL/HDL cholesterol were negatively correlated. The coronary risk factors which generally showed positive correlations with lipid parameters were BMI and blood pressure. Positive correlations were also recorded between fasting blood glucose and TG and uric acid with TG, TC/HDL cholesterol, and LDL/HDL cholesterol. In contrast, risk factors of negative correlations were observed between HDL cholesterol and the coronary risk factors of BMI, diastolic blood pressure, and uric acid. Smoking showed raised percent lipid abnormality for TG, HDL cholesterol, and TC/HDL cholesterol and LDL/HDL cholesterol. Alcohol consumption also increased the mean level and abnormality frequency for TG. The implication of this investigation is discussed.

Haffner et al. (1996) observed that the relation of possible metabolic precursors (especially insulin resistance) to hypertension has been controversial. In addition, these associations may differ by level of obesity or ethnicity. Age, body mass index, waist-to-hip ratio, and fasting insulin and triglyceride levels predicted the development of hypertension in univariate analyses. After adjustment for age, body mass index, waist-to-hip ratio, gender, ethnicity, and fasting glucose levels, higher levels of triglyceride and fasting insulin predicted the development of hypertension. Body mass index and fasting insulin and triglyceride levels predicted the development of hypertension in Mexican Americans and non-Hispanic whites. In addition, fasting insulin levels predicted the development of hypertension in lean and obese subjects. Increased insulin secretion (as judged by the 30-min insulin increment) on an oral glucose tolerance test also predicted the development of hypertension. A cluster of atherogenic changes may precede the development of hypertension, and increased fasting insulin concentration predicts hypertension in important subgroups of subjects.

Enas et al. (1996) compared the prevalence of coronary heart disease (CHD) and its risk factors in first-generation Asian Indian immigrants to the United States of

America (USA) with those of the native Caucasian population. The age-adjusted prevalence of myocardial infarction and/or angina was approximately three times more in Asian Indian men compared to the Framingham Offspring Study (7.2 % vs. 2.5 %; $P < 0.0001$) but was similar in women (0.3 % vs. 1 %; $P = 0.64$). Asian Indians had higher prevalence of non-insulin-dependent diabetes mellitus (NIDDM; 7.6 % vs. 1 %; $P < 0.0001$) but markedly lower prevalence of cigarette smoking (1.3 % vs. 27 %; $P < 0.0001$) and obesity (4.2 % vs. 22 %; $P < 0.0001$). Hypertension was less prevalent in Asian Indian men (14.2 % vs. 19.1 %, $P < 0.008$) but similar in women (11.3 % vs. 11.4 %). The prevalence of elevated total and low-density lipoprotein (LDL)-cholesterol levels was similar in men [17 % vs. 23.4 % ($P = 0.24$) and 13.7 % vs. 22.3 % ($P = 0.22$), respectively] but lower in women [15 % vs. 26.1 % ($P = 0.018$) and 14.3 % vs. 19.6 % ($P = 0.047$), respectively]. The mean levels of high-density lipoprotein (HDL) cholesterol were less in younger (30–39 years) Asian Indian men (mean: 0.98 vs. 1.18 mmol/L; $P < 0.001$) and middle-aged (30–59 years) women (mean: 1.24 vs. 1.45 mmol/L; $P < 0.001$). The prevalence of hypertriglyceridemia was similar in men (18.5 % vs. 11.3 %) but higher in Asian Indian women (8.3 % vs. 4.1 %, $P = 0.02$). To conclude, immigrant Asian Indian men to the USA have high prevalence of CHD, NIDDM, low HDL-cholesterol levels, and hypertriglyceridemia. All these have "insulin resistance" as a common pathogenetic mechanism and seem to be the most important risk factors.

Hokanson and Austin (1996) reported that despite nearly 40 years of research, the role of plasma triglyceride as a risk factor for cardiovascular disease remains elusive. The present study was to quantify the magnitude of the association between triglyceride and cardiovascular disease in the general population and determine whether this relationship is independent of high-density lipoprotein (HDL) cholesterol, using the semiquantitative techniques of meta-analysis. For men and women, the univariate RRs for triglyceride were 1.32 (95 % CI 1.26–1.39) and 1.76 (95 % CI 1.50–2.07), respectively, indicating an approximately 30 % increased risk in men and a 75 % increase in women. Adjustment of HDL cholesterol and other risk factors attenuated these RRs to 1.14 (95 % CI 1.05–1.28) and 1.37 (95 % CI 1.13–1.66), respectively, which were still statistically significant values. Based on combined data from prospective studies, triglyceride is a risk factor for cardiovascular disease for both men and women in the general population, independent of HDL cholesterol. These findings demonstrate the necessity for clinical trials to evaluate whether lowering plasma triglyceride decreases the risk of cardiovascular disease.

Whelton (1994) reported that cardiovascular disease is the main cause of death in virtually all industrialized countries. The limited information available from developing countries suggests that a similar epidemic is inevitable if current trends go unchecked. Treatment of patients with clinical manifestations is an important element in overall management but on its own is an insufficient and incomplete response. Sudden death is often the first manifestation of cardiovascular disease, and, even when treatment of disease is applicable and effective, it is usually palliative rather than curative. Thus, treatment and prevention directed at the underlying risk factors, including high blood pressure, constitute a complementary and more fundamental approach to reducing the burden of illness. Epidemiological studies provide the scientific foundation for such an approach by identifying the distribution and determinants of high blood pressure in the general population, by establishing the role of high blood pressure as a risk factor for cardiorenal complications and by quantifying the potential value of treating and preventing high blood pressure in the general population.

Newnham (1993) observed that cardiovascular disease remains the major cause of death for postmenopausal women in Western societies. The majority of epidemiological studies indicated that postmenopausal estrogen replacement therapy is associated with a 50 % reduction in the risk of cardiovascular disease, with much of the reduction being mediated by changes in the plasma concentration of cholesterol within high- and low-density lipoproteins. In addition to favorably influencing the plasma concentration of lipoproteins, estrogens also influence the complex metabolism of lipoproteins in the arterial wall, helping to impede the formation of the atherosclerotic plaque. Whilst estrogens alter endothelial function, vascular reactivity, and fibrinolysis, these changes are also seen with reduction of LDL cholesterol and may partly reflect the altered concentration of plasma lipoproteins induced by estrogens. Oral estrogens have substantially greater favorable effects on LDL and HDL cholesterol than their transdermal counterparts but also result in greater hypertriglyceridemia. Most progesterones antagonize the beneficial effects of estrogens on lipoproteins in a dose-dependent manner; however, cyclical use of low doses of progestogens with an oral estrogen generally retains a net beneficial effect. Lipoprotein levels fluctuate during cyclical therapy, the most adverse changes being noted at the end of the progesterone phase. Lipoprotein concentrations are constant during continuous combined regimens which have the potential for more prolonged exposure to an adverse progestational effect. Finally, estrogen therapy may be useful in the management of postmenopausal women with hyperlipidemia and also in the secondary prevention of clinical sequelae in women with established atherosclerosis.

Swai et al. (1993) assessed the prevalence of risk factors for coronary heart disease (CHD) in rural Tanzanians. Mean serum cholesterol levels in men were 4.2, 3.4, and 3.7 mmol/L and in women 4.4, 3.6, and 3.9 mmol/L in Kilimanjaro, Morogoro, and Mara regions, respectively. In Kilimanjaro region, 17.4 % of men and 19.0 % of women had values above 5.2 mmol/L compared with only 5.0 % and 6.7 % in Morogoro region and 4.8 % and 6.9 %, respectively, in Mara region. Systolic and diastolic blood pressures increased with age in both men and women in all three regions with the most marked increase in Kilimanjaro region and the smallest rise in Mara region. Mean age-adjusted values were highest in Kilimanjaro region (124/75 mmHg and 125/76 mmHg in men and women, respectively) and lowest in Mara region (120/70 mmHg in men and 118/68 mmHg in women). Hypertension was found in 6.6 % of men and 7.5 % of women in Kilimanjaro region, 3.3 % and 4.7 % in Morogoro, and 2.6 % and 3.4 % in Mara region. Cigarette smoking was found in 42.6 % of men in Kilimanjaro region, 28.2 % in Morogoro region, and 8.6 % in Mara region. Less than 4 % of women smoked in all three regions.

Hong et al. (1992) examined the effects of estrogen replacement on lipids and angiographically defined CAD in postmenopausal women; lipid profiles were obtained in 90 consecutive postmenopausal women undergoing diagnostic coronary angiography. 20 % were receiving estrogen and 80 % were not. CAD (defined as greater than or equal to 25 % luminal diameter narrowing in a major coronary artery) was present in only 22 % of women (4 of 18) receiving estrogen and in 68 % (49 of 72) who were not ($P < 0.001$), with an odds ratio of 0.13. Mean HDL-C level was significantly higher (63 ± 6 vs. 48 ± 2; $P < 0.01$) and mean total/HDL-cholesterol ratio significantly lower in women receiving estrogen than in those who were not (4.2 ± 0.5 vs. 5.1 ± 0.2; $P < 0.05$). The other lipid values were similar in both groups. On multiple logistic regression analysis, absence of estrogen use was the most powerful independent predictor of the presence of CAD ($P < 0.001$), with total/HDL-cholesterol ratio as the only other variable selected ($P < 0.01$). Thus, among 90 consecutive postmenopausal women undergoing diagnostic coronary angiography, estrogen replacement therapy was associated with an 87 % reduction in the prevalence of CAD, and those receiving estrogen had a significantly higher mean HDL cholesterol level and lower mean total/HDL-cholesterol ratio.

Tao et al. (1992) conducted surveys in four Chinese populations, urban and rural for both Beijing and Guangzhou, as part of PRC-USA collaborative research in cardiovascular and cardiopulmonary epidemiology. Age-adjusted mean serum TC was higher in urban than rural samples and generally higher in Beijing than Guangzhou, ranging from 155 mg/dL for Guangzhou rural women to 187 mg/dL for Guangzhou urban women. Group mean values of HDL cholesterol varied from 48 to 59 mg/dL, higher in Beijing than Guangzhou and higher in women than men. TC/HDL-cholesterol ratio ranged from 3.05 to 3.82. Serum TG values were higher for Beijing than Guangzhou; the lowest group mean values of 78 and 75 mg/dL were in rural Guangzhou men and women. Mean body mass index (BMI) was uniformly low, ranging from 20 kg/m^2 for rural Guangzhou to 24 kg/m^2 for urban Beijing. Multiple regression analyses showed that BMI was positively and independently related to serum TC, LDL cholesterol, TG, and TC/HDL cholesterol and inversely related to HDL cholesterol. Smoking was positively related in both sexes to TG and TC/HDL cholesterol and inversely related to HDL cholesterol. Smoking was also positively related to TC and LDL cholesterol in men. In men, alcohol was positively related to TC and HDL cholesterol and was inversely related to TG and TC/HDL cholesterol. Heavy manual work was inversely related to TC, LDL cholesterol, and TC/HDL cholesterol in men, but not related to lipids in women. Thus, for these Chinese population samples, despite their lower serum TC and BMI, the correlates of serum lipids are similar to those in Western populations. These variables accounted for only part of the observed urban–rural and north–south differences in serum lipids among these Chinese population samples. The significance of the relatively low serum TC and TG and high HDL cholesterol in relation to low cardiovascular disease in Chinese populations is the object of further investigation in follow-up studies.

Gordon et al. (1989) reported in the British Regional Heart Study (BRHS) that much of the inverse relation of high-density lipoprotein cholesterol (HDL-C) and incidence of coronary heart disease was eliminated by covariance adjustment. Using the proportional hazards model and adjusting for age, blood pressure, smoking, body mass index, and low-density lipoprotein cholesterol, they analyzed this relation separately in the Framingham Heart Study (FHS), Lipid Research Clinics Prevalence Mortality Follow-up Study (LRCF) and Coronary Primary Prevention Trial (CPPT), and Multiple Risk Factor Intervention Trial (MRFIT). A 1-mg/dL (0.026 mM) increment in HDLC was associated with a significant coronary heart disease risk decrement of 2 % in men (FHS, CPPT, and MRFIT) and 3 % in women (FHS). In LRCF, where only fatal outcomes were documented, a 1-mg/dL increment in HDLC was associated with significant 3.7 % (men) and 4.7 % (women) decrements in cardiovascular disease mortality rates. The 95 % confidence intervals for these decrements in coronary heart and cardiovascular disease risk in the four studies overlapped considerably, and all contained the range 1.9–2.9 %. HDL-C levels were essentially unrelated to noncardiovascular disease mortality. When differences in analytical methodology were eliminated, a consistent inverse relation of HDL-C levels and coronary heart disease event rates was apparent in BRHS as well as in the four American studies.

A prospective survey has been undertaken of a total community of men and women, aged 35–69 years at recruitment, living in Port-of-Spain, Trinidad, by **Beckles et al.** (1986). By comparison with adults of African descent, age-adjusted relative risks of death from all-causes and from cardiovascular diseases were significantly increased in those of Indian origin (1.5 and 2.6, respectively) and reduced in those of mixed descent (0.5 and 0.3, respectively). Adults of European descent had an all-cause and cardiovascular mortality relative risk of 0.8 and 2.1, respectively. These ethnic differences in risk were not explained by systolic blood pressure, fasting blood glucose concentration, serum high-density lipoprotein or low-density lipoprotein concentration, or smoking habits. Differences in risk of cardiovascular death between Indian and European men seemed to be accounted for by the high prevalence of diabetes in Indians (19 %), but other ethnic contrasts in mortality were unrelated to diabetes mellitus.

According to Ryder et al. (1984), because acute myocardial infarction may affect plasma lipid concentrations, it is commonly recommended that assessment of these concentrations should be delayed until about 3 months after the acute event. A study was therefore conducted of fasting plasma lipid concentrations in patients with acute myocardial infarction. Measurements were made during their stay in hospital (days 1, 2, and 9) and 3 months later. Triglyceride concentrations remained unchanged throughout. Values of total cholesterol, low-density lipoprotein, and high-density lipoprotein all fell significantly between the first 2 days and day 9. Total cholesterol and low-density lipoprotein also showed significant falls between days 1 and 2. Nevertheless, fasting plasma lipid concentrations showed no significant difference at any time during the first 48 h from values measured 3 months later. After the infarction, 26 patients changed to eating less fat or less energy, or both. More patients had hypercholesterolemia in the first 48 h than at 3 months. These results suggested that lipid state may be assessed as accurately and possibly more accurately, during the first 48 h after acute myocardial infarction than at 3 months.

Miller and Miller (1975) reported that the body cholesterol pool increases with decreasing plasma high-density lipoprotein (HDL) but is unrelated to the plasma concentrations of total cholesterol and other lipoproteins. This finding supported existing evidence that HDL facilitates the uptake of cholesterol from peripheral tissues and its transport to the liver for catabolism and excretion. Plasma HDL is reduced in several conditions associated with an increased risk of future ischemic heart disease (IHD), namely, hypercholesterolemia, hypertriglyceridemia, male sex, obesity, and diabetes mellitus, while subjects with existing clinical IHD have lower levels of HDL than healthy subjects within the same community. It was proposed that a reduction of plasma HDL concentration may accelerate the development of atherosclerosis and hence IHD by impairing the clearance of cholesterol from the arterial wall.

Vetter et al. (1974) measured plasma concentrations of certain substrates and hormones serially and hourly from 1 to 5 h after the onset of major symptoms in patients with acute myocardial infarction. Plasma total catecholamines, free fatty acids, cortisol, and glucose were increased within 1 h and probably within 30 min of the onset of symptoms. Plasma insulin concentrations were low initially and rose later to normal values. Plasma cyclic adenosine monophosphate concentrations were increased during the first 3 h. Plasma triglycerides fell progressively from the second hour. Plasma cholesterol remained unchanged. Knowledge of these early systemic responses to acute myocardial infarction in man should aid the rational development of treatment designed to maintain the viability of the myocardium at a time when ischemia and its consequences may be reversible.

References

American Heart Association (AHA) (2003) Heart disease and stroke statistics: 2004 update. American Heart Association, Dallas, TX

Ballantyne CM (1998) Low-density lipoproteins and risk for coronary artery disease. Am J Cardiol 82:3Q–12Q

Beckles GL, Miller GJ, Kirkwood BR, Alexis SD, Carson DC (1986) High total and cardiovascular disease mortality in adults of Indian descent in Trinidad, unexplained by major coronary risk factors. Lancet 1:1298–1301

Bermudez OI, Velez-Carrasco W, Schaefer EJ, Tucker KL (2002) Dietary and plasma lipid, lipoprotein, and apolipoprotein profiles among elderly Hispanics and non-Hispanics and their association with diabetes. Am J Clin Nutr 76(6):1214–1221

Burman A, Jain K, Gulati R, Chopra V, Agrawal DP, Vaisisht S (2004) Lipoprotein (a) as a marker of coronary artery disease and its association with dietary fat. J Assoc Physicians India 52:99–102

Ellison RC, Zhang Y, Qureshi MM, Knox S, Arnett DK, Province MA, Investigators of the NHLBI Family Heart Study (2004) Lifestyle determinants of high-density lipoprotein cholesterol: the National Heart, Lung, and Blood Institute Family Heart Study. Am Heart J 147:529–535

Enas EA, Garg A, Davidson MA (1996) Coronary heart disease and its risk factors in first-generation immigrant Asian Indians to the United States of America. Indian Heart J 48:343–352

Gaziano JM, Sesso HD, Breslow JL (1999) Relation between systemic hypertension and blood lipids on the risk of myocardial infarction. Am J Cardiol 84:768–773

Gordon DJ, Probstfield JL, Garrison RJ, Neaton JD, Castelli WP, Knoke JD (1989) High-density lipoprotein cholesterol and cardiovascular disease. Four prospective American studies. Circulation 79:8–15

Gould AL, Rossouw JE, Santanello NC, Heyse JF, Furberg CD (1998) Cholesterol reduction yields clinical benefit: impact of statin trials. Circulation 97:946–952

Gururajan P, Gurumurthy P, Nayar P, Chockalingam M, Bhuvaneshwari S, Babu S, Sarasabharati A, Victor D, Cherian KM (2010) Lipid profile and non-enzymic antioxidant status in patients with acute coronary syndrome in South India. Heart Lung Circ 19(2):75–80

Habib SS (2006) Frequency distribution of atherogenic dyslipidemia in Saudi type 2 diabetic patients. Pak J Physiol 2:20–23

Haffner SM, American Diabetes Association (2004) Dyslipidemia management in adults with diabetes. Diabetes Care 27:S68–S71

Haffner SM, Miettinen H, Gaskill SP, Stern MP (1996) Metabolic precursors of hypertension: the San Antonio Heart Study. Arch Intern Med 156:1994–2000

Hokanson JE, Austin MA (1996) Plasma triglyceride level is a risk factor for cardiovascular disease independent of high-density lipoprotein cholesterol level: a meta-analysis of population-based prospective studies. J Cardiovasc Risk 3:213–219

Hong MK, Romm PA, Reagan K, Green CE, Rackley CE (1992) Effects of oestrogen replacement therapy on serum lipid values and angiographically defined coronary artery disease in post-menopausal women. Am J Cardiol 69:176–178

Idemudia JO, Ugwuja EI (2009) Plasma lipid profiles in hypertensive Nigerians. Internet J Cardiovasc Res 6(2):8

Iqbal MP, Shafiq M, Mehboobali N, Iqbal SP, Abbasi K (2004) Variability in lipid profile in patients with acute myocardial infarction from two tertiary care hospitals in Pakistan. J Pak Med Assoc 54(11):544–549

Johnson ML, Pietz K, Battleman DS, Beyth RJ (2004) Prevalence of comorbid hypertension and dyslipidemia and associated cardiovascular disease. Am J Manag Care 10:926–932

Joint National Committee on Detection, Evaluation, and Treatment of High Blood Pressure (JNC VI) (1997) The sixth reports of the Joint National Committee on Detection, Evaluation, and Treatment of High Blood Pressure (JNC VI). Arch Intern Med 157:2413–2446

Kanaley JA, Sames C, Swisher L, Swick AG, Steppan CM, Feiglin D, Jaynes EB, Meyer RA, Weinstock RS (2001) Abdominal fat distribution in pre and post menopausal women. The impact of physical activity, age and menopausal status. Metabolism 50(8):976–982

Karthikeyan G, Teo KK, Islam S, McQueen MJ, Pais P, Wang X, Sato H, Lang CC, Sitthi-Amorn C, Pandey MR, Kazmi K, Sanderson JE, Yusuf S (2009) Lipid profile, plasma apolipoproteins, and risk of a first myocardial infarction among Asians: an analysis from the INTERHEART Study. J Am Coll Cardiol 53(3):244–253

Khoo KL, Tan H, Leiw YM (1997) Serum lipids and their relationship with other coronary risk factors in healthy subjects in a city clinic. Med J Malaysia 52:38–52

Kumar A, Sivakanesan R (2009) Serum lipid profile abnormality in predicting the risk of myocardial infarction in elderly normolipidaemic patients in south Asia: a case-controlled study. Internet J Altern Med 6:2

Mackay J, Mensah G (2004) Atlas of heart disease and stroke. World Health Organization, Geneva

Mengesha AY (2006) Lipid profile among diabetes patients in Gaborone, Botswana. S Afr Med J 96:147–148

Mgonda YM, Ramaiya KL, Swai ABM, Mc-Larty DG, George KM, Alberti M (1998) Insulin resistance and hypertension in non-obese Africans in Tanzania. Hypertension 31:114–118

Miller GJ, Miller NE (1975) Plasma high density lipoprotein concentration and development of ischemic heart disease. Lancet 1:16–19

Mosca L, Appel LJ, Benjamin EJ, Berra K, Chandra-Strobos N, Fabunmi RP, Grady D, Haan CK, Hayes SN, Judelson DR, Keenan NL, McBride P, Oparil S, Ouyang P, Oz MC, Mendelsohn

ME, Pasternak RC, Pinn VW, Robertson RM, Schenck-Gustafsson K, Sila CA, Smith SC Jr, Sopko G, Taylor AL, Walsh BW, Wenger NK, Williams CL, American Heart Association (2004) Evidence-based guidelines for cardiovascular disease prevention in women. Circulation 109:672–693

Newnham HH (1993) Oestrogens and atherosclerotic vascular disease-lipid factors. Bailleres Clin Endocrinol Metab 7:61–93

Nigam PK, Narain VS, Hasan M (2004) Serum lipid profile in patients with acute myocardial infarction. Indian J Clin Biochem 19(1):67–70

Njelekela M, Negishi H, Nara Y, Tomohiro M, Kuga S, Noguchi T, Kanda T, Yamori M, Mashalla Y, Jian Liu L (2001) Cardiovascular risk factors in Tanzania: a revisit. Acta Trop 79(3):231–239

Nthangeni G, Steyn NP, Alberts M (2002) Dietary intake and barriers to dietary compliance in black type II diabetic patients attending primary health-care services. Public Health Nutr 5:329–338

Oparil S, Zaman MA, Calhoun DA (2003) Pathogenesis of hypertension. Ann Intern Med 139:761–776

Ryder REJ, Hayes TM, Mulligan IP, Kingwood JC, Williams S, Owens DR (1984) How soon after myocardial infarction should plasma lipid values be assessed. Br Med J 289:1651–1653

Saaddine JB, Engelgau MM, Beckles GL (2002) A diabetes report card for the United States: quality of care in the 1990s. Ann Intern Med 136:565

Saha MS, Sana NK, Shaha RK (2006) Serum lipid profile of hypertensive patients in the Northern region of Bangladesh. J Biosci 14:93–98

Seyoum B, Abdulkadir J, Berhanu P (2003) Analysis of serum lipids and lipoproteins in Ethiopian diabetic patients. Ethiop Med J 41(1):1–8

Singh RB, Beegom R, Ghosh S, Niaz MA, Rastogi V (1997) Epidemiological study of hypertension and its determinants in an urban population of North India. J Hum Hypertens 11:679–685

Smith S, Lall AM (2008) A study on lipid profile levels of diabetics and non-diabetics among Naini region of Allahabad, India. Turk J Biochem 33(4):138–141

Swai AB, McLarty DG, Kitange HM, Kilima PM, Tatalla S, Keen N, Chuwa LM, Alberti KG (1993) Low prevalence of risk factors for coronary heart disease in rural Tanzania. Int J Epidemiol 22(4):651–659

Tao S, Li Y, Xiao Z, Cen R, Zhang H, Zhuo B, Chen P, Liao Y (1992) Serum lipids and their correlates in Chinese urban and rural population of Beijing and Guangzhou. PRC-USA Cardiovascular and Cardiopulmonary Epidemiology Research Group. J Epidemiol 21:893–903

Tsimikas S, Bergmark C, Beyer RW (2003) Temporal increases in plasma markers of oxidized low density lipoprotein strongly reflect the presence of acute coronary syndrome. J Am Coll Cardiol 41:360–370

Ugwu CE, Ezeanyika LUS, Daikwo MA, Amana R (2009) Lipid profile of a population of diabetic patients attending Nigerian National Petroleum Corporation Clinic, Abuja. Afr J Biochem Res 3(3):066–069

Vetter NJ, Strange RC, Adams W, Oliver MF (1974) Initial metabolic and hormonal response to acute myocardial infarction. Lancet 1(7852):284–289

Wenger NK (2003) Coronary heart disease: the female heart is vulnerable. Prog Cardiovasc Dis 46:199–229

Whelton PK (1994) Epidemiology of hypertension. Lancet 334:101–106

Wilsgaard T, Arnesen E (2004) Change in serum lipids and body mass index by age, sex, and smoking status: the Tromso study 1986–1995. Ann Epidemiol 14:265–273

Youmbissi TJ, Djoumessi S, Nouedoui C (2001) Profile lipidique d'un group d'hypertendus camerounais noir Africains. Med Afr Noire 31:114–118

Chapter 5
Materials and Methods

5.1 Study Samples

Study samples comprised of 800 patients/subjects and were selected from Punjab Institute of Cardiology, Lahore. One-half of them were males and another half were females. Each gender was classed into two age groups, i.e., above 50 years of age and below 50 years of age. Each subgroup of age was further spliced into four categories of heart disease or associated risk factors (angina, angina with coexistent diabetes, hypertension, and myocardial infarction). A group of 50 normal individuals was also included in each subgroup. The control subjects (healthy individuals: males and females) were selected from among apparently healthy subjects. Informed consent was obtained from all participants of the study. The study was approved by Ethical Committee of Institute of Chemistry, University of the Punjab, Lahore.

N.B. Rizvi and S.A. Nagra, *Minerals and Lipids Profiles in Cardiovascular Disorders in South Asia*, DOI 10.1007/978-3-642-34249-3_5,

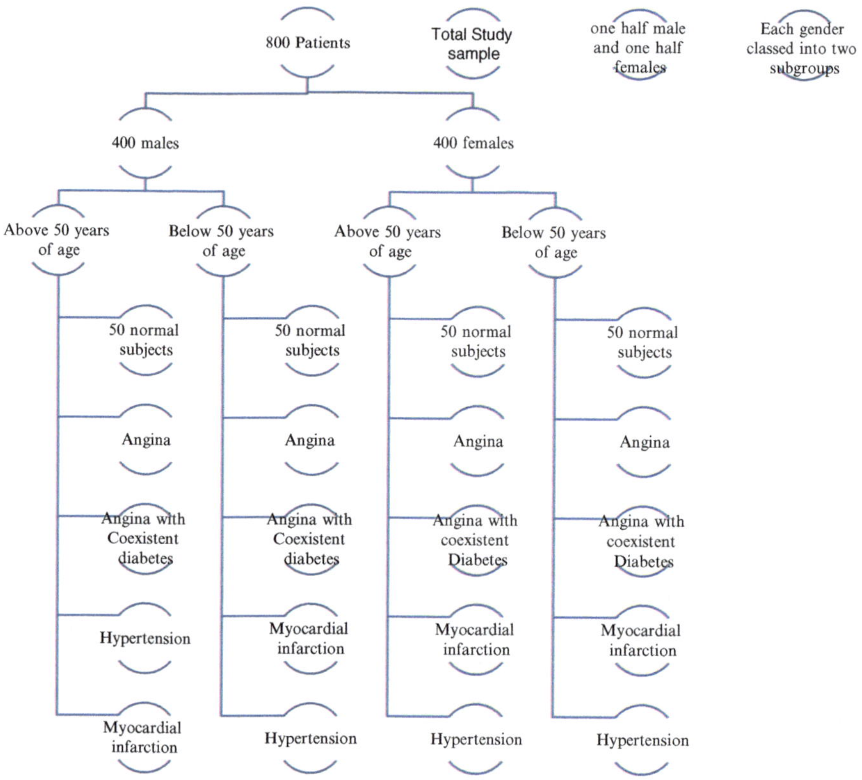

5.2 Inclusion Criteria

Only the subjects who came to the outdoor of the hospital the first time and any therapy had yet to be instituted were included in the study. Further they did not receive any vitamin or mineral supplement and had normal serum uric acid, creatinine, and urea level.

The detailed history about lifestyle (smoking, physical activity, dietary pattern), socioeconomic status, and personal and family history of cardiovascular diseases were also recorded on a special Performa with the consent of the patients and normal persons.

5.3 Basis of Age Grouping

To draw a cutoff line in age grouping is difficult. In the case of females, the comparison was between premenopausal and postmenopausal women. The average age for postmenopausal women falling in inclusion criteria was about 49.5 years and we rounded it off as 50 years. Available vital statistics of the hospital indicated that in the case of males, greater incidence of the CVD-related problems was encountered around 50 years. That formed the basis for classifying the data into two age groups in the case of male patients.

5.4 Criteria for Various Cardiovascular Disorders

Angina cases were characterized by chest pain radiating towards the back, sweating with shortness of breath, and with a remarkable change in their electrocardiogram (ECG).

Hypertension was defined as systolic blood pressure of 140 mmHg or higher and diastolic blood pressure of 90 mmHg or higher.

Myocardial infarction cases were identified by selecting indoor patients admitted to coronary care unit or equivalent cardiology ward, within 24 h of the onset of the problem.

Controls were categorized as normal healthy individuals without any history of heart disease or related problems.

5.5 Blood Sampling

Ten milliliters of fasting venous blood was collected with disposable syringe and then transferred to heparinized vacutainers, which were prechilled. The vacutainers were placed on ice till the plasma was separated. Plasma was prepared within 30 min of sample collection. Blood samples were centrifuged at 4,000 rpm for 10 min, using prechilled centrifuge tubes. Plasma was separated and stored at −20 °C until analysis. Plasma samples were analyzed in duplicates. All glassware used for the analysis were previously soaked in diluted nitric acid (10 %) and rinsed thoroughly with deionized water in order to avoid any contamination with trace metals. Openings of the glassware were covered with aluminum foil.

5.6 Analytical Procedure

5.6.1 Determination of Minerals

Copper, selenium, and zinc were analyzed in plasma using inductively coupled plasma optical emission spectrometer (ICP-OES). Magnesium was determined by flame atomic absorption spectrophotometer (AAS).

Standards for each element were also used to evaluate the results. The accuracy of the procedure was evaluated by analyzing commercially available samples of lyophilized human serum (Rockland Immunochemicals, USA).

5.6.2 Multielement Standard

A multielement standard US EPA 23 obtained from GFS Chemicals, containing each element to the concentration of 100 ppm, was used for the accurate analysis of trace minerals, i.e., copper, magnesium, selenium, and zinc. Traces of nonhazardous tartaric acid were also present in it.

5.6.3 Calibration Standards

Calibration standards for copper, magnesium, selenium, and zinc were prepared by making successive dilutions of the multielement standard (100 ppm) using deionized water.

For copper, selenium, and zinc, 1,000 ppb stock solution was prepared by diluting 1 mL of multielement standard (100 ppm) solution to 100 mL of deionized water. Further dilutions were made by using the following relationship:

$$C_1V_1 = C_2V_2 \tag{5.1}$$

$$V_1 = C_2V_2/C_1 \tag{5.2}$$

where, C_1 is the concentration of stock solution, V_1 is the volume of stock solution, C_2 is the concentration of required standard solution, and V_2 is the volume of standard solution.

Using Eq. (5.2), 50 ppb, 100 ppb, 150 ppb, and 200 ppb standards were prepared by diluting 5 mL, 10 mL, 15 mL, and 20 mL of stock (1,000 ppb) solution to 100 mL, respectively.

For magnesium, using relation (5.2), 2 ppm, 4 ppm, 6 ppm, 8 ppm, and 10 ppm solutions were prepared by diluting 1 mL, 2 mL, 3 mL, 4 mL, and 5 mL of multielement standard (100 ppm) to 50 mL using deionized water, respectively.

5.6.4 Preparation of Plasma Samples for Mineral Analyses

The plasma was thawed at room temperature half an hour before further analyses. One milliliter of the plasma was diluted with 4 mL of redistilled deionized water to make a total volume of 5 mL. This diluted plasma sample was filtered with a micropore membrane (Micropore Technologies, UK) with a pore size of 0.45 μm and a diameter of 25 mm. The filtration was done to remove turbidity or any suspensions present in the plasma sample that can interfere in analysis.

5.7 Instrumentation

5.7.1 *Inductively Coupled Plasma Optical Emission Spectroscopy*

Standard solutions were run one by one and reading was recorded. Diluted and filtered plasma samples were run one by one.

Inductively coupled plasma atomic emission spectroscopy (ICP-AES) also referred to as inductively coupled plasma optical emission spectroscopy (ICP-OES) is an analytical technique used for the qualitative and quantitative determination of trace metals and certain nonmetals in solution (PerkinElmer Optima 8x00 series).

5.7.1.1 Working Principle

It is a type of emission spectroscopy that uses the inductively coupled plasma to produce excited atoms and ions that emit electromagnetic radiation at wavelength characteristics of a particular element. The intensity of this emission is indicative of the concentration of the element within the sample. To generate plasma, first, argon gas is supplied to torch coil, and high-frequency electric current is applied to the work coil at the tip of the torch tube. Using the electromagnetic field created in the torch tube by the high-frequency current, argon gas is ionized and plasma is generated. This plasma has high electron density and temperature (8,000–10,000 K), and this energy is used for the excitation–emission of the sample. The basic working principle includes the introduction of sample solution into the core of inductively coupled argon plasma (ICP). At high plasma temperature, all elements become thermally excited and emit electromagnetic radiations at their characteristic wavelengths. These radiations are collected by the spectrometer and passes through a diffraction grating that serves to resolve the radiations into a spectrum of its constituent wavelengths. Within the spectrometer, this diffracted radiation is then collected by wavelength and amplified to yield an intensity measurement that can be converted to an elemental concentration by comparison with calibration standards. When plasma energy is given to an analyte from outside, the component elements (atoms) get excited. When the excited atoms return to ground state, emission rays (spectrum rays) are released and the emission rays that correspond to the photon wavelength are measured. The element type is determined by wavelength of the emitted rays, and the content of each element determination is based on the intensity of rays.

5.7.1.2 Working Procedure

Calibration line for each element (copper, selenium, and zinc) was obtained by measuring emission intensity of all standards one by one. Emission intensity for sample was also measured and its quantification was done by comparing it with calibration standards using calibration line.

5.7.2 Atomic Absorption Spectrophotometer

5.7.2.1 Working Principle

The technique makes use of absorption spectrometry to assess the concentration of an analyte in a sample. It relies therefore heavily on Beer–Lambert law.

> In Short, the electrons of the atoms in the atomizer can be promoted to higher orbital for an instant by absorbing a set quantity of energy (i.e. light of a given wavelength). This amount of energy (or wavelength) is specific to a particular electron transition in a particular element, and in general, each wavelength corresponds to only one element. This gives the technique its elemental selectivity.

As the quantity of energy (the power) put into the flame is known, and the quantity remaining at the other side (at the detector) can be measured, it is possible, from Beer–Lambert law, to calculate how many of these transitions took place and thus get a signal that is proportional to the concentration of the element being measured.

5.7.2.2 Working Procedure

Calibration line for magnesium was obtained by measuring absorption intensity of all standards one by one at selected wavelength. Absorption intensity for sample was also measured and its quantification was done by comparing it with calibration standards using calibration line.

5.8 Lipid Profile

In lipid profile, total cholesterol (TC), high-density lipoprotein cholesterol (HDL cholesterol), low-density lipoprotein cholesterol (LDL cholesterol), and triglycerides (TG) were estimated.

5.8.1 Total Cholesterol

Total cholesterol was determined by enzymatic calorimetric method using commercially available kit obtained from Merck (Germany).

$$\text{Cholesterol ester} + H_2O \xrightarrow{\text{Cholesterol esterase}} \text{Cholesterol} + \text{Fatty acid}$$

$$\text{Cholesterol} + O_2 \xrightarrow{\text{Cholesterol oxidase}} \text{Cholesterol-3-one} + H_2O_2$$

$$2H_2O_2 + \text{4-aminoantipyrine} + \text{phenol} \xrightarrow{\text{Peroxidase}} \text{Quinoneimine} + 4H_2O$$

5.8.1.1 Working Principle

Determination of cholesterol after enzymatic hydrolysis and oxidation. The colorimetric indicator is quinoneimine which is generated from 4-aminoantipyrine and phenol by hydrogen peroxide under the catalytic action of peroxidase.

5.8.1.2 Procedure

The test tubes were marked as blank, unknown, and standard. Ten microliters sample was added in sample test tube, 10 μL redistilled was added only in blank, and 10 μL standard was added in standard test tube. One milliliter of working reagent was added in each tube. All the tubes were incubated for 10 min at 37 °C. Absorbance of standard and samples was measured against the reagent blank at 500 nm.

$$\text{Cholesterol (mg/dL)} = A_{\text{sample}}/A_{\text{standard}} \times \text{Concentration of standard}$$

5.8.2 High-Density Lipoprotein Cholesterol

HDL cholesterol was determined by using commercially available kit used for total cholesterol and obtained from Merck Germany.

5.8.2.1 Working Principle

Chylomicrons, very-low-density lipoprotein cholesterol (VLDL), and LDL cholesterol were precipitated by adding phosphotungstic acid and magnesium ions to the plasma sample. Centrifugation left only the HDL cholesterol in the supernatant. Their cholesterol content was determined enzymatically using the reagent for total cholesterol.

5.8.2.2 Procedure

In centrifuge tube, 500 μL of precipitation reagent was added. Two hundred microliters of plasma sample was also added and let it stand for 10–15 min at room temperature. The mixture was centrifuged at 4,000 rpm (1,968 × *g*) for 10 min. A clear supernatant was obtained after centrifugation. Three test tubes were marked as blank, standard, and sample. Twenty microliters of the supernatant was added in 1 mL of the working reagent. In standard test tube, 20 μL of standard was added, while in the test tube labeled as blank, only working reagent was added. Mixed and incubated for 10 min at room temperature. Absorbance of samples and standard was read against reagent blank at 546 nm.

$$\text{HDL cholesterol (mg/dL)} = A_{\text{sample}}/A_{\text{standard}} \times \text{Concentration of standard}$$

5.8.3 *Low-Density Lipoprotein Cholesterol*

Low-density lipoprotein cholesterol (LDL cholesterol) was calculated by using the Friedewald et al. (1972) formula:

$$\text{LDL cholesterol} = (\text{Total cholesterol}) - (\text{HDL cholesterol}) - (\text{Triglyceride}/5)$$

5.8.4 *Triglyceride*

Triglycerides were determined by using commercially available kit used for total cholesterol and obtained from Merck Germany.

5.8.4.1 Working Principle

Determination of triglycerides after enzymatic splitting with lipoprotein lipase. Indicator is quinoneimine which is generated from 4-aminoantipyrine and 4-chlorophenol by hydrogen peroxide under the catalytic action of peroxidase.

$$\text{Triglycerides} + H_2O \xrightarrow{\text{Lipases}} \text{Glycerol} + \text{Fatty acids}$$

$$\text{Glycerol} + \text{ATP} \xrightarrow{\text{GK}} \text{Glycerol-3-phosphate} + \text{ADP}$$

$$\text{Glycerol-3-phosphate} + O_2 \xrightarrow{\text{GPO}} \text{Dihydroxy acetone phosphate} + H_2O_2$$

$$2H_2O_2 + \text{Aminoantipyrine} + \text{4-chlorophenol} \xrightarrow{\text{POD}} \text{Quinoneimine} + 4H_2O + \text{HCL}$$

5.8.4.2 Procedure

Three test tubes as blank, sample, and standard were marked. Ten microliters sample was added in sample test tube, 10 μL of standard was added in the standard test tube, and 10 μL of redistilled water was added into the third test tube to serve as blank. One milliliter of working reagent was then added in each test tube. All tubes were incubated at 37 °C for 5 min. Absorbance of unknown and standard was measured against the reagent blank at 546 nm.

$$\text{Triglyceride (mg/dL)} = A_{\text{sample}}/A_{\text{standard}} \times \text{Concentration of standard}$$

5.9 Statistical Analysis

Statistical analysis was carried out using SPSS software, version 13. The Levene test was used to test the assumption of homogeneity in order to apply the analysis of variance technique to assess the difference among the factors. Data were analyzed for the effect of age, gender, and various disease conditions on serum minerals, and lipid profile, using one-way ANOVA. Post hoc analyses were carried out using Fischer PLSD test. Various descriptive statistics were also worked out (Kennedy and Bush 1985).

References

Friedewald WT, Levy RI, Fredrickson DS (1972) Estimation of the concentration of low-density lipoprotein cholesterol in plasma, without use of the preparative ultracentrifuge. Clin Chem 18:499–502

Kennedy JJ, Bush AJ (1985) An introduction to the design and analysis of experiments in behavioral research. University Press of America, Lanham, MD

Chapter 6
Results: Minerals

6.1 Copper

The descriptive statistics regarding the serum copper concentration in different study groups has been summarized in Table 6.1.

The analysis of variance indicated a significant ($P < 0.05$) effect of age and gender on serum copper of all the study groups except control group (Tables 6.2, 6.3, and 6.4).

Multiple comparisons of mean serum concentration of copper in hypertensive patients belonging to different age and gender have been given in Table 6.5. Age did not affect the serum copper level in hypertensive males. Below 50 years gender did not appear to affect the copper concentration in serum. Similarly the serum copper level did not differ significantly in males above 50 years and females less than 50 years. All other comparisons were statistically significant ($P < 0.05$).

Multiple comparisons of mean serum concentration of copper in patients with myocardial infarction belonging to different age and gender have been given in Table 6.6. Below 50 years of age gender did not appear to affect the copper concentration in serum. All other comparisons were statistically significant ($P < 0.05$). *t*-Test was applied for comparison of serum copper level in different study groups and control subjects (Table 6.7). Serum copper values were significantly ($P < 0.05$) high in almost all age groups, gender, and all CV disorders when compared with the control.

6.2 Magnesium

The descriptive statistics regarding the serum copper concentration in different study groups has been summarized in Table 6.8.

The data were subjected to analysis of variance (Table 6.9) which indicated a significant ($P < 0.05$) effect of age and gender within all the study groups and control.

N.B. Rizvi and S.A. Nagra, *Minerals and Lipids Profiles in Cardiovascular Disorders in South Asia*, DOI 10.1007/978-3-642-34249-3_6,

Table 6.1 Descriptive statistics of serum copper (μg/dL) in different study groups as influenced by age and gender

Study group	Gender and age	Mean	Standard deviation	95 % confidence interval for mean	
				Lower	Upper
Angina	Male below 50	69.3	41.7	53.7	84.9
	Male above 50	107.3	40.3	92.2	122.3
	Female below 50	99.7	31.2	88.1	111.4
	Female above 50	122.4	37.6	108.3	136.5
Angina with coexistent diabetes mellitus	Male below 50	50.3	29.9	39.1	61.5
	Male above 50	122.0	35.9	108.5	135.4
	Female below 50	82.1	31.3	70.4	93.8
	Female above 50	125.5	39.5	110.7	140.2
Hypertension	Male below 50	96.5	27.9	86.0	106.9
	Male above 50	114.2	43.7	97.8	130.5
	Female below 50	103.9	35.5	90.6	117.2
	Female above 50	139.5	37.5	125.4	153.5
Myocardial infarction	Male below 50	63.6	38.8	49.1	78.1
	Male above 50	95.2	37.9	81.0	109.4
	Female below 50	59.2	27.8	48.8	69.6
	Female above 50	114.9	30.2	103.6	126.1
Control	Male below 50	44.5	22.1	36.2	52.8
	Male above 50	44.0	20.2	36.4	51.5
	Female below 50	58.4	25.8	48.7	68.0
	Female above 50	55.8	33.0	43.4	68.1

Normal range 65–165 μg/dL [Nagra et al (1991) Nutr Res 11:1341–1345]

Table 6.2 Analysis of variance indicating the effect of age and gender on serum copper concentration within different study groups and control

Study groups	*F*-ratio	Significance level
Angina	10.40	0.000*
Angina with coexistent diabetes mellitus	32.31	0.000*
Hypertension	7.86	0.000*
Myocardial infarction	18.14	0.000*
Control	2.52	0.061^{NS}

NS nonsignificant
*Significant at $P < 0.05$

Table 6.3 Multiple comparisons of mean serum copper (μg/dL) in angina patients belonging to different age and gender

Category (I)	Category (J)	(I − J)	Standard error	Significance
Male below 50	Male above 50	−38.0	9.79	0.000*
Male below 50	Female below 50	−30.4	9.79	0.002*
Male below 50	Female above 50	−53.1	9.79	0.000*
Male above 50	Female below 50	7.56	9.79	0.442^{NS}
Male above 50	Female above 50	−15.1	9.79	0.125^{NS}
Female below 50	Female above 50	−22.7	9.79	0.022*

NS nonsignificant
*Significant at $P < 0.05$

Table 6.4 Multiple comparisons of mean serum copper (μg/dL) in angina patients with coexistent DM belonging to different age and gender

Category (I)	Category (J)	(I − J)	Standard error	Significance
Male below 50	Male above 50	−71.6	8.87	0.000*
Male below 50	Female below 50	−31.8	8.87	0.000*
Male below 50	Female above 50	−75.1	8.87	0.000*
Male above 50	Female below 50	39.8	8.87	0.000*
Male above 50	Female above 50	−3.50	8.87	0.694^{NS}
Female below 50	Female above 50	−43.3	8.87	0.000*

NS nonsignificant
*Significant at $P < 0.05$

Table 6.5 Multiple comparisons of mean serum copper (μg/dL) in hypertensive patients belonging to different age and gender

Category (I)	Category (J)	(I − J)	Standard error	Significance
Male below 50	Male above 50	−17.6	9.46	0.064^{NS}
Male below 50	Female below 50	−7.41	9.46	0.435^{NS}
Male below 50	Female above 50	−42.9	9.46	0.000*
Male above 50	Female below 50	10.2	9.46	0.280^{NS}
Male above 50	Female above 50	−25.3	9.46	0.009*
Female below 50	Female above 50	−35.5	9.46	0.000*

NS nonsignificant
*Significant at $P < 0.05$

Table 6.6 Multiple comparisons of mean serum copper (μg/dL) in patients with myocardial infarction belonging to different age and gender

Category (I)	Category (J)	(I − J)	Standard error	Significance
Male below 50	Male above 50	−31.5	8.79	0.000*
Male below 50	Female below 50	4.44	8.79	0.614^{NS}
Male below 50	Female above 50	−51.2	8.79	0.000*
Male above 50	Female below 50	36.0	8.79	0.000*
Male above 50	Female above 50	−19.6	8.79	0.028*
Female below 50	Female above 50	55.6	8.79	0.000*

NS nonsignificant
*Significant at $P < 0.05$

Multiple comparisons of mean serum concentration of magnesium in angina patients belonging to different age and gender have been given in Table 6.10. In this study group, age did not affect the magnesium level in either gender. Similarly, effect of gender was nonsignificant in age group below 50 years.

Multiple comparisons of mean serum concentration of magnesium in angina patients with coexistent DM, and hypertensive patients belonging to different age and gender also followed the similar trend (Tables 6.11 and 6.12, respectively). However, multiple comparisons of mean serum magnesium level in MI did not

Table 6.7 Comparison of serum copper concentration in different study groups with control using *t*-test

	Angina	Angina + DM	Hypertension	Myocardial infarction
	P-value	*P*-value	*P*-value	*P*-value
Female	0.000*	0.000*	0.000*	0.000*
Male	0.000*	0.000*	0.000*	0.000*
Female < 50	0.000*	0.002*	0.000*	0.905
Female > 50	0.000*	0.000*	0.000*	0.000*
Male < 50	0.006*	0.398	0.000*	0.023*
Male > 50	0.000*	0.000*	0.000*	0.000*
<50	0.000*	0.008*	0.000*	0.067
>50	0.000*	0.000*	0.000*	0.000*
Overall	0.000*	0.000*	0.000*	0.000*

*Significant at $P < 0.05$

Table 6.8 Descriptive statistics of serum magnesium (mg/dL) in different study groups as influenced by age and gender

Study group	Gender and age	Mean	Standard deviation	95 % confidence interval for mean	
				Lower	Upper
Angina	Male below 50	1.41	0.227	1.33	1.50
	Male above 50	1.40	0.217	1.31	1.48
	Female below 50	1.52	0.220	1.44	1.60
	Female above 50	1.55	0.260	1.45	1.65
Angina with coexistent diabetes mellitus	Male below 50	1.10	0.379	0.963	1.24
	Male above 50	1.23	0.273	1.13	1.33
	Female below 50	1.31	0.181	1.24	1.38
	Female above 50	1.45	0.259	1.35	1.54
Hypertension	Male below 50	1.44	0.353	1.31	1.57
	Male above 50	1.50	0.306	1.39	1.62
	Female below 50	1.62	0.180	1.55	1.69
	Female above 50	1.72	0.315	1.61	1.84
Myocardial infarction	Male below 50	1.50	0.190	1.43	1.57
	Male above 50	1.61	0.302	1.49	1.72
	Female below 50	1.70	0.254	1.60	1.79
	Female above 50	1.61	0.294	1.50	1.72
Control	Male below 50	1.72	0.368	1.58	1.86
	Male above 50	1.74	0.365	1.60	1.88
	Female below 50	1.87	0.409	1.71	2.02
	Female above 50	1.96	0.326	1.84	2.08

Normal range 1.4–3.6 mg/dL [Nagra et al (1991) Nutr Res 11:1341–1345]

show an appreciable effect of age and gender (Table 6.13). Only in below 50 years that the effect of gender was statistically significant ($P < 0.05$)

Multiple comparisons of mean serum concentration of magnesium in control group belonging to different age and gender have been given in Table 6.14. On an

Table 6.9 Analysis of variance indicating the effect of age and gender on serum magnesium concentration within different study groups and control

Study groups	F-ratio	Significance level
Angina	3.24	0.024*
Angina with coexistent diabetes mellitus	7.95	0.000*
Hypertension	5.40	0.002*
Myocardial infarction	2.95	0.035*
Control	2.74	0.046*

*Significant at $P < 0.05$

Table 6.10 Multiple comparisons of mean serum magnesium (mg/dL) in angina patients belonging to different age and gender

Category (I)	Category (J)	(I − J)	Standard error	Significance
Male below 50	Male above 50	0.0153	0.059	0.799^{NS}
Male below 50	Female below 50	−0.107	0.059	0.076^{NS}
Male below 50	Female above 50	−0.137	0.059	0.024*
Male above 50	Female below 50	−0.122	0.059	0.043*
Male above 50	Female above 50	−0.153	0.059	0.012*
Female below 50	Female above 50	−0.0303	0.059	0.614^{NS}

NS nonsignificant
*Significant at $P < 0.05$

Table 6.11 Multiple comparisons of mean serum magnesium (mg/dL) in angina patients with coexistent DM belonging to different age and gender

Category (I)	Category (J)	(I − J)	Standard error	Significance
Male below 50	Male above 50	−0.128	0.072	0.080^{NS}
Male below 50	Female below 50	−0.208	0.072	0.005*
Male below 50	Female above 50	−0.347	0.072	0.000*
Male above 50	Female below 50	−0.079	0.072	0.277^{NS}
Male above 50	Female above 50	−0.218	0.072	0.003*
Female below 50	Female above 50	−0.138	0.072	0.060^{NS}

NS nonsignificant
*Significant at $P < 0.05$

Table 6.12 Multiple comparisons of mean serum magnesium (mg/dL) in hypertensive patients belonging to different age and gender

Category (I)	Category (J)	(I − J)	Standard error	Significance
Male below 50	Male above 50	−0.064	0.076	0.402^{NS}
Male below 50	Female below 50	−0.182	0.076	0.019*
Male below 50	Female above 50	−0.283	0.076	0.000*
Male above 50	Female below 50	−0.117	0.076	0.127^{NS}
Male above 50	Female above 50	−0.219	0.076	0.005*
Female below 50	Female above 50	−0.101	0.076	0.187^{NS}

NS nonsignificant
*Significant at $P < 0.05$

Table 6.13 Multiple comparisons of mean serum magnesium (mg/dL) in MI patients belonging to different age and gender

Category (I)	Category (J)	(I − J)	Standard error	Significance
Male below 50	Male above 50	−0.109	0.068	0.111^{NS}
Male below 50	Female below 50	−0.202	0.068	0.004*
Male below 50	Female above 50	−0.116	0.068	0.090^{NS}
Male above 50	Female below 50	−0.092	0.068	0.177^{NS}
Male above 50	Female above 50	−0.007	0.068	0.918^{NS}
Female below 50	Female above 50	0.085	0.068	0.212^{NS}

NS nonsignificant
*Significant at $P < 0.05$

Table 6.14 Multiple comparisons of mean serum magnesium (mg/dL) in control group belonging to different age and gender

Category (I)	Category (J)	(I − J)	Standard error	Significance
Male below 50	Male above 50	−0.017	0.095	0.856^{NS}
Male below 50	Female below 50	−0.128	0.095	0.130^{NS}
Male below 50	Female above 50	−0.235	0.095	0.015*
Male above 50	Female below 50	−0.128	0.095	0.181^{NS}
Male above 50	Female above 50	−0.218*	0.095	0.024*
Female below 50	Female above 50	−0.090	0.095	0.345^{NS}

NS nonsignificant
*Significant at $P < 0.05$

Table 6.15 Comparison of serum magnesium concentration in different study groups with control using *t*-test

	Angina	Angina + DM	Hypertension	MI
	P-value	*P*-value	*P*-value	*P*-value
Female	0.000*	0.000*	0.000*	0.002*
Male	0.000*	0.000*	0.000*	0.000*
Female < 50	0.000*	0.000*	0.000*	0.000*
Female > 50	0.000*	0.000*	0.009*	0.129^{NS}
Male < 50	0.000*	0.000*	0.004*	0.060
Male > 50	0.000*	0.000*	0.006*	0.000*
<50	0.000*	0.000*	0.000*	0.001*
>50	0.000*	0.000*	0.000*	0.000*
Overall	0.000*	0.000*	0.000*	0.000*

NS nonsignificant
*Significant at $P < 0.05$

average, females less than 50 years of age had significantly ($P < 0.05$) higher serum magnesium than males of either age group.

t-Test was applied for comparison of serum magnesium level in different study groups and control subjects (Table 6.15). Serum magnesium values were significantly ($P < 0.05$) lower in almost all age groups, gender, and all CV disorders when compared with the control.

Table 6.16 Descriptive statistics of serum selenium (μg/dL) in different study groups as influenced by age and gender

Study group	Gender and age	Mean	Standard deviation	95 % confidence interval for mean	
				Lower	Upper
Angina	Male below 50	78.32	19.79	70.93	85.71
	Male above 50	83.04	20.65	75.32	90.75
	Female below 50	76.49	23.55	67.69	85.28
	Female above 50	83.65	20.37	76.04	91.25
Angina with coexistent diabetes mellitus	Male below 50	75.48	14.90	69.91	81.05
	Male above 50	79.92	20.65	72.21	87.63
	Female below 50	58.37	14.60	52.92	63.83
	Female above 50	87.64	18.97	80.55	94.72
Hypertension	Male below 50	80.70	18.69	73.71	87.68
	Male above 50	80.32	21.82	72.17	88.47
	Female below 50	84.17	20.56	76.49	91.85
	Female above 50	82.17	18.51	75.26	89.08
Myocardial infarction	Male below 50	75.10	15.53	69.30	80.90
	Male above 50	80.52	20.01	73.04	87.99
	Female below 50	74.11	17.60	67.54	80.68
	Female above 50	83.45	17.28	77.00	89.91
Control	Male below 50	70.32	18.83	63.29	77.35
	Male above 50	76.54	29.58	65.49	87.59
	Female below 50	74.74	19.07	67.61	81.86
	Female above 50	103.90	27.39	93.67	114.12

Normal range 50–100 μg/dL [Vasudevan DM, Sreekumari S (2001) Text book of biochemistry. Jaypee Br., New Delhi]

Table 6.17 Analysis of variance indicating the effect of age and gender on serum selenium concentration within different study groups and control

Study groups	*F*-ratio	Significance level
Angina	0.830	0.480^{NS}
Angina with coexistent diabetes mellitus	15.05	0.000*
Hypertension	0.231	0.875^{NS}
Myocardial infarction	1.89	0.13^{NS}
Control	11.88	0.000*

NS nonsignificant
*Significant at $P < 0.05$

6.3 Selenium

The descriptive statistics regarding the serum selenium concentration in different study groups has been summarized in Table 6.16.

The analysis of variance indicated a significant ($P < 0.05$) effect of age and gender on serum selenium of angina patients with coexistent DM and control group. A nonsignificant effect of age and gender was observed on serum selenium in the

Table 6.18 Multiple comparisons of mean serum selenium (μg/dL) in angina patients with coexistent DM belonging to different age and gender

Category (I)	Category (J)	(I − J)	Standard error	Significance
Male below 50	Male above 50	−4.440	4.513	0.327[NS]
Male below 50	Female below 50	17.106*	4.513	0.000*
Male below 50	Female above 50	−12.156*	4.513	0.008*
Male above 50	Female below 50	21.546*	4.513	0.000*
Male above 50	Female above 50	−7.716	4.513	0.090[NS]
Female below 50	Female above 50	−29.263*	4.513	0.000*

NS nonsignificant
*Significant at $P < 0.05$

Table 6.19 Multiple comparisons of mean serum selenium (μg/dL) in control group belonging to different age and gender

Category (I)	Category (J)	(I − J)	Standard error	Significance
Male below 50	Male above 50	−6.220	6.251	0.322[NS]
Male below 50	Female below 50	−4.416	6.251	0.481[NS]
Male below 50	Female above 50	−33.573*	6.251	0.000*
Male above 50	Female below 50	1.803	6.251	0.773[NS]
Male above 50	Female above 50	−27.353*	6.251	0.000*
Female below 50	Female above 50	−29.156*	6.251	0.000*

NS nonsignificant
*Significant at $P < 0.05$

Table 6.20 Comparison of serum selenium in different study groups with control using *t*-test

	Angina	Angina + DM	Hypertension	Myocardial infarction
	P-value	*P*-value	*P*-value	*P*-value
Female	0.045*	0.001*	0.161[NS]	0.015[NS]
Male	0.082[NS]	0.283[NS]	0.089[NS]	0.270[NS]
Female < 50	0.753[NS]	0.000*	0.071[NS]	0.895[NS]
Female > 50	0.158[NS]	0.184[NS]	0.065[NS]	0.084[NS]
Male < 50	0.114[NS]	0.244[NS]	0.036*	0.288[NS]
Male > 50	0.328[NS]	0.610[NS]	0.576[NS]	0.544[NS]
Age < 50	0.191[NS]	0.090[NS]	0.006*	0.523[NS]
Age > 50	0.158[NS]	0.184[NS]	0.065[NS]	0.084[NS]
Overall	0.751[NS]	0.054[NS]	0.880[NS]	0.303[NS]

NS nonsignificant
*Significant at $P < 0.05$

case of patients with angina alone, hypertension, and myocardial infarction (Table 6.17).

Multiple comparisons of means of serum selenium in patients having angina with coexistent DM and control group have been summarized in Table 6.18. The effect of age on serum selenium was observed nonsignificant in the case of males. Similarly effect of gender on serum selenium was also nonsignificant in below

Table 6.21 Descriptive statistics of serum zinc (μg/dL) in different study groups as influenced by age and gender

Study group	Gender and age	Mean	Standard deviation	95 % confidence interval for mean	
				Lower	Upper
Angina	Male below 50	24.18	18.68	17.20	31.15
	Male above 50	19.58	19.66	12.24	26.62
	Female below 50	41.26	25.72	31.65	50.87
	Female above 50	31.86	21.13	23.97	39.75
Angina with coexistent diabetes mellitus	Male below 50	29.68	21.52	21.64	37.72
	Male above 50	24.70	19.50	17.41	31.98
	Female below 50	25.63	15.28	19.92	31.34
	Female above 50	24.16	16.96	17.83	30.50
Hypertension	Male below 50	23.03	14.44	17.63	28.42
	Male above 50	27.66	23.83	18.76	36.56
	Female below 50	45.50	29.48	34.48	56.51
	Female above 50	28.28	18.83	21.25	35.31
Myocardial infarction	Male below 50	25.33	16.69	19.09	31.56
	Male above 50	20.63	20.26	13.06	28.19
	Female below 50	33.41	26.49	23.52	43.30
	Female above 50	34.65	26.95	24.58	44.71
Control	Male below 50	116.6	73.53	89.20	144.12
	Male above 50	90.50	49.86	71.87	109.12
	Female below 50	78.33	55.62	57.56	99.10
	Female above 50	96.25	67.55	71.02	121.47

Normal range 80–198 μg/dL [Nagra et al (1991) Nutr Res 11:1341–1345]

Table 6.22 Analysis of variance for zinc levels in control and study groups belonging to different gender and age groups

Study groups	*F*-ratio	Significance level
Angina	5.86	0.001*
Angina with coexistent diabetes mellitus	0.549	0.650^{NS}
Hypertension	5.83	0.001*
Myocardial infarction	2.52	0.061^{NS}
Control	1.97	0.121^{NS}

NS nonsignificant
*Significant at $P < 0.05$

50 years of age. Multiple comparisons of means of serum selenium in control group have been given in Table 6.19. *t*-Test was applied for comparison of serum selenium level in different study groups and control subjects (Table 6.20). Strikingly, no appreciable difference was observed in control and different study groups.

Females and male less than 50 years of age having angina with coexistent DM and hypertension, respectively, had significantly lower serum selenium than control group. All other comparisons were observed nonsignificant.

Table 6.23 Multiple comparisons of mean serum zinc (μg/dL) in angina patients belonging to different age and gender

Category (I)	Category (J)	(I − J)	Standard error	Significance
Male below 50	Male above 50	4.60	5.54	0.408[NS]
Male below 50	Female below 50	−17.08	5.54	0.003*
Male below 50	Female above 50	−7.68	5.54	0.168[NS]
Male above 50	Female below 50	−21.68	5.54	0.000*
Male above 50	Female above 50	−12.28	5.54	0.029*
Female below 50	Female above 50	9.40	5.54	0.093[NS]

NS nonsignificant
*Significant at $P < 0.05$

Table 6.24 Multiple comparisons of mean serum (μg/dL) zinc in hypertensive patients belonging to different age and gender

Category (I)	Category (J)	(I − J)	Standard error	Significance
Male below 50	Male above 50	−4.63	5.77	0.424[NS]
Male below 50	Female below 50	−22.46	5.77	0.000*
Male below 50	Female above 50	−5.25	5.77	0.365[NS]
Male above 50	Female below 50	−17.83	5.77	0.003*
Male above 50	Female above 50	−0.616	5.77	0.915[NS]
Female below 50	Female above 50	17.21	5.77	0.004*

NS nonsignificant
*Significant at $P < 0.05$

Table 6.25 Comparison of serum zinc in different study groups with control using *t*-test

	Angina	Angina + DM	Hypertension	MI
Study groups	*P*-value	*P*-value	*P*-value	*P*-value
Female	0.000*	0.000*	0.000*	0.000*
Male	0.000*	0.000*	0.000*	0.000*
Female < 50	0.002*	0.000*	0.006*	0.000*
Female > 50	0.000*	0.000*	0.000*	0.000*
Male < 50	0.000*	0.000*	0.000*	0.000*
Male > 50	0.000*	0.000*	0.000*	0.000*
<50	0.000*	0.000*	0.000*	0.000*
>50	0.000*	0.000*	0.000*	0.000*
Overall	0.000*	0.000*	0.000*	0.000*

*Significant at $P < 0.05$

6.4 Zinc

The descriptive statistics regarding the serum zinc concentration in different study groups has been summarized in Table 6.21. The data were subjected to analysis of variance (Table 6.22) which indicated a significant ($P < 0.05$) effect of age and gender within angina and hypertensive patients.

Multiple comparisons of mean serum concentration of zinc in angina patients belonging to different age and gender have been summarized in Table 6.23. Males below 50 years of age had significantly ($P < 0.05$) lower serum zinc than females of the same age group. Males above 50 years of age had significantly ($P < 0.05$) lower serum zinc than females of either age.

Multiple comparisons of mean serum concentration of zinc in hypertensive patients belonging to different age and gender have been given in Table 6.24. Effect of age was nonsignificant in the case of males. However, female above 50 years of age had significantly ($P < 0.05$) high serum zinc concentration than below 50 years of age.

t-Test was applied for comparison of serum zinc level in different study groups and control subjects (Table 6.25). Serum zinc values were significantly ($P < 0.05$) lower in almost all age groups, gender, and all CV disorders when compared with the control.

Chapter 7
Results: Lipid Profile

7.1 Total Cholesterol

The descriptive statistics regarding the serum total cholesterol in different study groups has been summarized in Table 7.1. The data were subjected to analysis of variance (Table 7.2) which indicated a significant effect of gender and age on all the study groups.

Multiple comparisons of mean cholesterol level in angina patients (Table 7.3) indicated a significant effect ($P < 0.05$) of age on total cholesterol level of both the genders. Females less than 50 years of age had significantly ($P < 0.05$) lower serum total cholesterol level than their male counterparts. Likewise, females above 50 years of age had significantly higher ($P < 0.05$) serum total cholesterol than their male counterparts.

Multiple comparisons of mean cholesterol level in angina patients with coexistent diabetes mellitus (Table 7.4) indicated a nonsignificant effect of age on total cholesterol level of males and a significant effect ($P < 0.05$) on females. Females above 50 years of age had significantly ($P < 0.05$) higher serum total cholesterol level than males below 50 years of age. All other comparisons were nonsignificant statistically.

Multiple comparisons of mean cholesterol level in hypertensive patients (Table 7.5) indicated a nonsignificant effect of age on total cholesterol level of males and a significant effect ($P < 0.05$) on females. Females below 50 years of age had significantly ($P < 0.05$) lower serum total cholesterol level than males below 50 years of age. Similarly females above 50 had significantly ($P < 0.05$) higher serum total cholesterol level than males above 50 years of age. All other comparisons were nonsignificant statistically.

Multiple comparisons of mean cholesterol level in patients of myocardial infarction (Table 7.6) indicated a nonsignificant effect of age on total cholesterol level of males and a significant effect ($P < 0.05$) on females. Females above 50 years of age had significantly ($P < 0.05$) higher serum total cholesterol level than males of either age group. All other comparisons were nonsignificant statistically. In the case

N.B. Rizvi and S.A. Nagra, *Minerals and Lipids Profiles in Cardiovascular Disorders in South Asia*, DOI 10.1007/978-3-642-34249-3_7,

Table 7.1 Descriptive statistics of total cholesterol (mg/dL) in different study groups as influenced by age and gender

Study group	Gender and age	Mean	Standard deviation	95 % confidence interval for mean Lower	Upper
Angina	Male below 50	184.93	53.25	165.04	204.81
	Male above 50	161.06	42.99	145.01	177.12
	Female below 50	145.00	29.89	133.83	156.16
	Female above 50	187.00	31.76	175.13	198.86
Angina with coexistent diabetes mellitus	Male below 50	178.76	33.07	166.41	191.11
	Male above 50	180.50	36.78	166.76	194.23
	Female below 50	173.00	32.58	160.83	185.16
	Female above 50	196.83	31.91	184.91	208.74
Hypertension	Male below 50	188.40	36.74	174.67	202.12
	Male above 50	174.03	26.79	164.02	184.03
	Female below 50	162.90	35.94	149.47	176.32
	Female above 50	192.13	32.30	180.06	204.19
Myocardial infarction	Male below 50	167.96	31.04	156.37	179.55
	Male above 50	160.53	33.68	147.95	173.11
	Female below 50	165.46	42.98	149.41	181.51
	Female above 50	200.00	22.96	191.42	208.57
Control	Male below 50	123.23	41.58	107.70	138.76
	Male above 50	137.86	40.79	122.63	153.10
	Female below 50	119.66	25.05	110.31	129.02
	Female above 50	145.36	32.49	133.23	157.50

Normal range < 200 mg/dL (NCEP 2001)

Table 7.2 Analysis of variance for total cholesterol level in control and study groups belonging to different gender and age groups

Study groups	*F*-ratio	Significance level
Angina	7.38	0.000*
Angina with coexistent diabetes mellitus	2.77	0.045*
Hypertension	4.94	0.003*
Myocardial infarction	8.63	0.000*
Control	3.47	0.018*

*Significant at $P < 0.05$

Table 7.3 Multiple comparisons of total cholesterol (mg/dL) level in angina patients belonging to different age and gender

Category (I)	Category (J)	(I – J)	Standard error	Significance
Male below 50	Male above 50	23.86	10.47	0.025*
Male below 50	Female below 50	39.93	10.47	0.000*
Male below 50	Female above 50	−2.066	10.47	0.844^{NS}
Male above 50	Female below 50	16.06	10.47	0.128^{NS}
Male above 50	Female above 50	−25.93	10.47	0.015*
Female below 50	Female above 50	−42.00	10.47	0.000*

NS nonsignificant

*Significant at $P < 0.05$

Table 7.4 Multiple comparisons of total cholesterol (mg/dL) level in angina patients with coexistent diabetes mellitus belonging to different age and gender

Category (I)	Category (J)	(I − J)	Standard error	Significance
Male below 50	Male above 50	−1.73	8.68	0.842^{NS}
Male below 50	Female below 50	5.76	8.68	0.508^{NS}
Male below 50	Female above 50	−18.06	8.68	0.40*
Male above 50	Female below 50	7.50	8.68	0.390^{NS}
Male above 50	Female above 50	−16.33	8.68	0.063^{NS}
Female below 50	Female above 50	−23.83	8.68	0.007*

NS nonsignificant
*Significant at $P < 0.05$

Table 7.5 Multiple comparisons of total cholesterol (mg/dL) level in hypertensive patients belonging to different age and gender

Category (I)	Category (J)	(I − J)	Standard error	Significance
Male below 50	Male above 50	14.36	8.56	0.096^{NS}
Male below 50	Female below 50	25.50	8.56	0.004*
Male below 50	Female above 50	−3.73	8.56	0.664^{NS}
Male above 50	Female below 50	11.13	8.56	0.196^{NS}
Male above 50	Female above 50	−18.10	8.56	0.037*
Female below 50	Female above 50	−29.23	8.56	0.001*

NS nonsignificant
*Significant at $P < 0.05$

Table 7.6 Multiple comparisons of total cholesterol (mg/dL) level in patients of myocardial infarction belonging to different age and gender

Category (I)	Category (J)	(I − J)	Standard error	Significance
Male below 50	Male above 50	7.43	8.63	0.391^{NS}
Male below 50	Female below 50	2.50	8.63	0.773^{NS}
Male below 50	Female above 50	−32.03	8.63	0.000*
Male above 50	Female below 50	−4.93	8.63	0.569^{NS}
Male above 50	Female above 50	−39.46	8.63	0.000*
Female below 50	Female above 50	−34.53	8.63	0.000*

NS nonsignificant
*Significant at $P < 0.05$

of control group, multiple comparisons of mean indicated that females above 50 years of age had significantly high ($P < 0.05$) level of serum total cholesterol than males and females below 50 years of age (Table 7.7).

A comparison of total cholesterol level between control and other study groups has been summarized in Table 7.8. It was observed that all the study groups had significantly higher ($P < 0.05$) serum total cholesterol level than the control. A comparison of total cholesterol level between control and other study groups has been summarized in Table 7.8.

It was observed that all the study groups had significantly higher ($P < 0.05$) serum total cholesterol level than the control.

Table 7.7 Multiple comparisons of means of total cholesterol (mg/dL) in control subjects belonging to different age and gender

Category (I)	Category (J)	(I − J)	Standard error	Significance
Male below 50	Male above 50	−14.63	9.19	0.114^{NS}
Male below 50	Female below 50	3.56	9.19	0.699^{NS}
Male below 50	Female above 50	−22.13*	9.19	0.018*
Male above 50	Female below 50	18.20	9.19	0.050^{NS}
Male above 50	Female above 50	−7.50	9.19	0.417^{NS}
Female below 50	Female above 50	−25.70*	9.19	0.006*

NS nonsignificant
*Significant at $P < 0.05$

Table 7.8 Comparison of serum total cholesterol level in different study groups with control using *t*-test

	Angina	Angina + DM	Hypertension	MI
	P-value	*P*-value	*P*-value	*P*-value
Female	0.000*	0.000*	0.000*	0.000*
Male	0.000*	0.000*	0.000*	0.000*
Female < 50	0.001*	0.000*	0.000*	0.000*
Female > 50	0.000*	0.000*	0.000*	0.000*
Male < 50	0.000*	0.000*	0.000*	0.000*
Male > 50	0.036*	0.000*	0.000*	0.022*
<50	0.000*	0.000*	0.000*	0.000*
>50	0.000*	0.000*	0.000*	0.000*
Overall	0.000*	0.000*	0.000*	0.000*

*Significant at $P < 0.05$

7.2 High-Density Lipoprotein Cholesterol

The descriptive statistics regarding the serum high-density lipoprotein cholesterol (HDL cholesterol) in different study groups has been summarized in Table 7.9.

The data were subjected to analysis of variance (Table 7.10) which indicated a significant ($P < 0.05$) effect of gender and age on HDL cholesterol in angina patients with coexistent diabetes mellitus. There was a nonsignificant effect of age and gender on other study groups.

Multiple comparisons of mean HDL-cholesterol level in angina patients with coexistent diabetes mellitus (Table 7.11) indicated a significant effect ($P < 0.05$) of age on HDL-cholesterol level of males only. Males below 50 years had significantly lower ($P < 0.05$) HDL cholesterol than females of both the age groups.

A comparison of HDL-cholesterol level between control and other study groups has been summarized in Table 7.12. It was observed that dominantly it were the patients of myocardial infarction who had significantly lower ($P < 0.05$) serum HDL-cholesterol level than the control.

Table 7.9 Descriptive statistics of HDL cholesterol (mg/dL) in different study groups as influenced by age and gender

Study group	Gender and age	Mean	Standard deviation	95 % confidence interval for mean	
				Lower	Upper
Angina	Male below 50	34.46	11.70	30.09	38.83
	Male above 50	34.96	9.44	31.43	38.49
	Female below 50	35.70	9.72	32.06	39.33
	Female above 50	30.06	10.66	26.08	34.04
Angina with coexistent diabetes mellitus	Male below 50	34.66	10.83	30.61	38.71
	Male above 50	26.13	9.55	22.56	29.70
	Female below 50	33.36	9.64	29.76	36.96
	Female above 50	31.33	9.33	27.84	34.82
Hypertension	Male below 50	36.46	11.55	32.15	40.78
	Male above 50	31.06	11.38	26.81	35.31
	Female below 50	35.36	9.74	31.72	39.00
	Female above 50	32.33	10.30	28.48	36.18
Myocardial infarction	Male below 50	31.63	7.49	28.83	34.43
	Male above 50	31.93	10.04	28.18	35.68
	Female below 50	32.73	9.46	29.19	36.26
	Female above 50	30.60	10.05	26.84	34.35
Control	Male below 50	38.06	10.74	34.05	42.07
	Male above 50	38.83	10.61	34.86	42.79
	Female below 50	36.83	9.18	33.40	40.26
	Female above 50	35.60	10.99	31.49	39.70

Normal range > 35 mg/dL (NCEP 2001)

Table 7.10 Analysis of variance for HDL-cholesterol level in control and study groups belonging to different gender and age groups

Study groups	F-ratio	Significance level
Angina	1.78	0.154^{NS}
Angina + DM	4.34	0.006*
Hypertension	1.65	0.181^{NS}
Myocardial infarction	0.268	0.848^{NS}
Control	0.557	0.644^{NS}

NS nonsignificant
*Significant at $P < 0.05$

7.3 Low-Density Lipoprotein Cholesterol

The descriptive statistics regarding the serum total cholesterol in different study groups has been summarized in Table 7.13.

The data were subjected to analysis of variance (Table 7.14) which indicated a significant effect of gender and age on all the study groups except control group.

Multiple comparisons of mean low-density lipoprotein-cholesterol (LDL-cholesterol) level in angina patients (Table 7.15) indicated a significant effect

Table 7.11 Multiple comparisons of HDL-cholesterol (mg/dL) level in angina patients with coexistent diabetes mellitus belonging to different age and gender

Category (I)	Category (J)	(I − J)	Standard error	Significance
Male below 50	Male above 50	8.53	2.54	0.001*
Male below 50	Female below 50	1.30	2.54	0.611[NS]
Male below 50	Female above 50	3.33	2.54	0.193[NS]
Male above 50	Female below 50	−7.23	2.54	0.005*
Male above 50	Female above 50	−5.20	2.54	0.043*
Female below 50	Female above 50	2.03	2.54	0.426[NS]

NS nonsignificant
*Significant at $P < 0.05$

Table 7.12 Comparison of serum HDL-cholesterol level in different study groups with control using *t*-test

	Angina	Angina + DM	Hypertension	Myocardial infarction
	P-value	*P*-value	*P*-value	*P*-value
Female	0.079[NS]	0.032*	0.200[NS]	0.013*
Male	0.056[NS]	0.000*	0.023*	0.000*
Female < 50	0.644[NS]	0.159[NS]	0.551[NS]	0.094[NS]
Female > 50	0.053[NS]	0.111[NS]	0.240[NS]	0.071[NS]
Male < 50	0.220[NS]	0.227[NS]	0.581[NS]	0.009*
Male > 50	0.142[NS]	0.000*	0.008*	0.012*
<50	0.211[NS]	0.064[NS]	0.415[NS]	0.002*
>50	0.016*	0.000*	0.006*	0.002*
Overall	0.009*	0.000*	0.011*	0.000*

NS nonsignificant
*Significant at $P < 0.05$

($P < 0.05$) of age on LDL-cholesterol level of females only. Males below 50 years had significantly higher ($P < 0.05$) LDL cholesterol than females of the same age. Females above 50 years of age had significantly ($P < 0.05$) higher LDL-cholesterol level than all the study groups.

Multiple comparisons of LDL-cholesterol level in angina patients with coexistent diabetes mellitus revealed that females above 50 years of age had significantly ($P < 0.05$) high LDL cholesterol than males of either age group (Table 7.16).

Hypertensive females above 50 years of age had significantly ($P < 0.05$) higher LDL cholesterol than males above 50 and females below 50 years of age. Males below 50 years had significantly ($P < 0.05$) higher LDL cholesterol than females of the same age (Table 7.17). Almost similar trend of variation in LDL cholesterol was observed in patients of myocardial infarction (Table 7.18).

A comparison of LDL-cholesterol level between control and other study groups has been summarized in Table 7.19. It was observed that all the study groups had significantly higher ($P < 0.05$) serum LDL-cholesterol level than the control.

Table 7.13 Descriptive statistics of LDL cholesterol (mg/dL) in different study groups as influenced by age and gender

Study group	Gender and age	Mean	Standard deviation	95 % confidence interval for mean	
				Lower	Upper
Angina	Male below 50	116.23	59.08	94.17	138.29
	Male above 50	97.46	40.82	82.22	112.70
	Female below 50	85.03	31.34	73.32	96.73
	Female above 50	125.06	35.34	111.86	114.13
Angina with coexistent diabetes mellitus	Male below 50	106.10	33.93	93.42	118.77
	Male above 50	98.06	32.59	85.89	110.23
	Female below 50	113.36	31.71	101.52	125.20
	Female above 50	127.23	32.84	114.96	139.49
Hypertension	Male below 50	121.16	40.92	105.88	136.44
	Male above 50	106.83	26.59	96.90	116.76
	Female below 50	102.06	36.44	88.45	115.67
	Female above 50	127.46	33.51	114.95	139.98
Myocardial infarction	Male below 50	94.40	32.65	82.20	106.59
	Male above 50	100.20	35.75	86.85	113.54
	Female below 50	100.26	41.67	84.70	115.82
	Female above 50	136.70	29.99	125.49	147.90
Control	Male below 50	60.56	40.16	45.56	75.56
	Male above 50	75.13	36.40	61.54	88.72
	Female below 50	63.50	24.71	54.27	72.72
	Female above 50	80.43	30.90	68.89	91.97

Normal range < 130 mg/dL (NCEP 2001)

Table 7.14 Analysis of variance for LDL-cholesterol level in control and study groups belonging to different gender and age groups

Study groups	*F*-ratio	Significance level
Angina	5.30	0.002*
Angina + DM	4.28	0.007*
Hypertension	3.52	0.017*
Myocardial infarction	9.06	0.000*
Control	2.36	0.075^{NS}

NS nonsignificant
*Significant at $P < 0.05$

Table 7.15 Multiple comparisons of LDL-cholesterol (mg/dL) level in angina patients belonging to different age and gender

Category (I)	Category (J)	(I − J)	Standard error	Significance
Male below 50	Male above 50	18.76	11.09	0.094^{NS}
Male below 50	Female below 50	31.20	11.09	0.006*
Male below 50	Female above 50	−8.83	11.09	0.428^{NS}
Male above 50	Female below 50	12.43	11.09	0.265^{NS}
Male above 50	Female above 50	−27.60	11.09	0.014*
Female below 50	Female above 50	−40.03	11.09	0.000*

NS nonsignificant
*Significant at $P < 0.05$

Table 7.16 Multiple comparisons of LDL cholesterol (mg/dL) level in angina patients with coexistent diabetes mellitus belonging to different age and gender

Category (I)	Category (J)	(I − J)	Standard error	Significance
Male below 50	Male above 50	8.03	8.46	0.345^{NS}
Male below 50	Female below 50	−7.26	8.46	0.392^{NS}
Male below 50	Female above 50	−21.13	8.46	0.014*
Male above 50	Female below 50	−15.30	8.46	0.073^{NS}
Male above 50	Female above 50	−29.16	8.46	0.001*
Female below 50	Female above 50	−13.86	8.46	0.104^{NS}

NS nonsignificant
*Significant at $P < 0.05$

Table 7.17 Multiple comparisons of LDL-cholesterol (mg/dL) level in hypertensive patients belonging to different age and gender

Category (I)	Category (J)	(I − J)	Standard error	Significance
Male below 50	Male above 50	14.33	8.97	0.113^{NS}
Male below 50	Female below 50	19.10	8.97	0.035*
Male below 50	Female above 50	−6.30	8.97	0.484^{NS}
Male above 50	Female below 50	4.76	8.97	0.596^{NS}
Male above 50	Female above 50	−20.63	8.97	0.023*
Female below 50	Female above 50	−25.40	8.97	0.005*

NS nonsignificant
*Significant at $P < 0.05$

Table 7.18 Multiple comparisons of LDL-cholesterol (mg/dL) level in patients of myocardial infarction belonging to different age and gender

Category (I)	Category (J)	(I − J)	Standard error	Significance
Male below 50	Male above 50	−5.80	9.11	0.526^{NS}
Male below 50	Female below 50	−5.86	9.11	0.521^{NS}
Male below 50	Female above 50	−42.30	9.11	0.000*
Male above 50	Female below 50	−0.066	9.11	0.994^{NS}
Male above 50	Female above 50	−36.50	9.11	0.000*
Female below 50	Female above 50	−36.43	9.11	0.000*

NS nonsignificant
*Significant at $P < 0.05$

7.4 Triglycerides

The descriptive statistics regarding the serum triglycerides in different study groups belonging to different gender and age has been summarized in Table 7.20.

The data were subjected to analysis of variance (Table 7.21) which indicated a significant ($P < 0.05$) effect of gender and age on all the study groups.

Multiple comparisons of mean serum triglyceride level in angina patients (Table 7.22) indicated a significant effect ($P < 0.05$) of age on serum triglyceride

Table 7.19 Comparison of serum LDL-cholesterol level in different study groups with control using *t*-test

	Angina	Angina + DM	Hypertension	Myocardial infarction
	P-value	*P*-value	*P*-value	*P*-value
Female	0.000*	0.000*	0.000*	0.000*
Male	0.000*	0.000*	0.000*	0.000*
Female < 50	0.005*	0.000*	0.000*	0.000*
Female > 50	0.000*	0.000*	0.000*	0.000*
Male < 50	0.000*	0.000*	0.000*	0.001*
Male > 50	0.029*	0.013*	0.000*	0.009*
<50	0.000*	0.000*	0.000*	0.000*
>50	0.000*	0.000*	0.000*	0.000*
Overall	0.000*	0.000*	0.000*	0.000*

*Significant at $P < 0.05$

Table 7.20 Descriptive statistics of serum triglycerides (mg/dL) in different study groups as influenced by age and gender

Study group	Gender and age	Mean	Standard deviation	95 % confidence interval for mean Lower	Upper
Angina	Male below 50	170.90	78.34	141.64	200.15
	Male above 50	143.70	53.58	123.69	163.70
	Female below 50	121.13	49.28	102.72	139.53
	Female above 50	159.00	40.85	143.74	174.25
Angina with coexistent diabetes mellitus	Male below 50	190.00	32.85	177.73	202.26
	Male above 50	285.36	144.17	231.53	339.20
	Female below 50	131.53	41.53	116.02	147.04
	Female above 50	191.16	72.01	164.27	218.05
Hypertension	Male below 50	154.53	39.04	139.95	169.11
	Male above 50	181.16	51.32	162.00	200.33
	Female below 50	127.33	29.85	116.18	138.48
	Female above 50	161.00	37.25	147.08	174.91
Myocardial infarction	Male below 50	209.73	79.66	179.98	239.47
	Male above 50	142.46	46.80	124.98	159.94
	Female below 50	160.10	39.58	145.31	174.88
	Female above 50	160.66	33.30	148.23	173.10
Control	Male below 50	123.16	45.11	106.31	140.01
	Male above 50	119.76	36.15	106.26	133.26
	Female below 50	96.26	18.79	89.24	103.28
	Female above 50	146.76	35.65	133.45	160.07

Normal range < 200 mg/dL (NCEP 2001)

of females only. Males below 50 years had significantly higher ($P < 0.05$) serum triglyceride than females of the same age.

Multiple comparisons of serum triglyceride level in angina patients with coexistent diabetes mellitus belonging to different age and gender have been summarized

Table 7.21 Analysis of variance for serum triglyceride level in control and study groups belonging to different gender and age groups

Study groups	F-ratio	Significance level
Angina	4.22	0.007*
Angina + DM	16.89	0.000*
Hypertension	9.20	0.000*
Myocardial infarction	8.95	0.000*
Control	10.31	0.000*

*Significant at $P < 0.05$

Table 7.22 Multiple comparisons of serum triglyceride (mg/dL) level in angina patients belonging to different age and gender

Category (I)	Category (J)	(I – J)	Standard error	Significance
Male below 50	Male above 50	27.20	14.78	0.068^{NS}
Male below 50	Female below 50	49.76	14.78	0.001*
Male below 50	Female above 50	11.90	14.78	0.422^{NS}
Male above 50	Female below 50	22.56	14.78	0.130^{NS}
Male above 50	Female above 50	−15.30	14.78	0.303^{NS}
Female below 50	Female above 50	−37.86	14.78	0.012*

NS nonsignificant
*Significant at $P < 0.05$

Table 7.23 Multiple comparisons of serum triglyceride level (mg/dL) in angina patients with coexistent diabetes mellitus belonging to different age and gender

Category (I)	Category (J)	(I – J)	Standard error	Significance
Male below 50	Male above 50	−95.36	21.89	0.000*
Male below 50	Female below 50	58.46	21.89	0.009*
Male below 50	Female above 50	−1.16	21.89	0.958^{NS}
Male above 50	Female below 50	153.83	21.89	0.000*
Male above 50	Female above 50	94.20	21.89	0.000*
Female below 50	Female above 50	−59.63	21.89	0.007*

NS nonsignificant
*Significant at $P < 0.05$

in Table 7.23. Effect of age and gender appear to be significant ($P < 0.05$) on all comparisons except males below 50 years and females above 50 years of age. Males above 50 had the significantly ($P < 0.05$) highest serum triglycerides than all other groups.

Multiple comparisons of serum triglyceride level in hypertensive patients belonging to different age and gender have been summarized in Table 7.24. Effect of age and gender appear to be significant ($P < 0.05$) on all comparisons except that males of either age had significantly ($P < 0.05$) higher triglycerides than females of below 50 years.

Table 7.24 Multiple comparisons of serum triglyceride level (mg/dL) in hypertensive patients belonging to different age and gender

Category (I)	Category (J)	(I − J)	Standard error	Significance
Male below 50	Male above 50	−26.63	10.35	0.011*
Male below 50	Female below 50	27.20	10.35	0.010*
Male below 50	Female above 50	−6.46	10.35	0.534^NS^
Male above 50	Female below 50	53.83	10.35	0.000*
Male above 50	Female above 50	20.16	10.35	0.054^NS^
Female below 50	Female above 50	−33.66	10.35	0.002*

NS nonsignificant
*Significant at $P < 0.05$

Table 7.25 Multiple comparisons of serum triglyceride level (mg/dL) in patients of myocardial infarction belonging to different age and gender

Category (I)	Category (J)	(I − J)	Standard error	Significance
Male below 50	Male above 50	67.26	13.67	0.000*
Male below 50	Female below 50	49.63	13.67	0.000*
Male below 50	Female above 50	49.06	13.67	0.000*
Male above 50	Female below 50	−17.63	13.67	0.200^NS^
Male above 50	Female above 50	−18.20	13.67	0.186^NS^
Female below 50	Female above 50	−0.566	13.67	0.967^NS^

NS nonsignificant
*Significant at $P < 0.05$

Table 7.26 Multiple comparisons of serum triglyceride level (mg/dL) in control group belonging to different age and gender

Category (I)	Category (J)	(I − J)	Standard error	Significance
Male below 50	Male above 50	3.40	9.09	0.709^NS^
Male below 50	Female below 50	26.90	9.09	0.004*
Male below 50	Female above 50	−23.60	9.09	0.011*
Male above 50	Female below 50	23.50	9.09	0.011*
Male above 50	Female above 50	−27.00	9.09	0.004*
Female below 50	Female above 50	−50.500	9.09	0.000*

NS nonsignificant
*Significant at $P < 0.05$

Multiple comparisons of mean of serum triglycerides in patients of myocardial infarction (Table 7.25) belonging to different age groups and gender indicated a significant ($P < 0.05$) effect of age on serum triglyceride level of males. However, the effect of age on serum triglyceride level of females was observed as nonsignificant. Females of either age group had significantly lower serum triglyceride than males below 50 years of age.

In the case of control group, multiple comparisons of serum triglyceride level indicated a nonsignificant effect of age on triglyceride level of males. Females

Table 7.27 Comparison of serum triglyceride level in different study groups with control using *t*-test

	Angina	Angina + DM	Hypertension	Myocardial infarction
	P-value	*P*-value	*P*-value	*P*-value
Female	0.022*	0.000*	0.001*	0.000*
Male	0.001*	0.000*	0.000*	0.000*
Female < 50	0.014*	0.000*	0.000*	0.000*
Female > 50	0.222[NS]	0.004*	0.136[NS]	0.124[NS]
Male < 50	0.005*	0.000*	0.006*	0.000*
Male > 50	0.047*	0.000*	0.000*	0.040*
<50	0.001*	0.000*	0.000*	0.000*
>50	0.024*	0.000*	0.000*	0.013*
Overall	0.000*	0.000*	0.000*	0.000*

NS nonsignificant
*Significant at $P < 0.05$

Table 7.28 Percentage of patients having different CV disorders and falling in different ranges

Lipid profile parameters	Range (mg/dL)	Angina	Angina + DM	Hypertension	Myocardial infarction	Control
Cholesterol	<200	77.5	68.3	71.7	75.0	95.0
	≥200	22.5	31.7	28.3	25.0	5.0
HDL cholesterol	<35	48.3	54.2	50.8	55.0	34.2
	35–65	51.7	45.8	49.2	45.0	65.8
	>65	0.0	0.0	0.0	0.0	0.0
LDL cholesterol	<130	70.0	73.3	62.5	70.0	93.3
	≥130	30.0	26.7	37.5	30.0	6.7
Triglyceride	<200	86.3	64.2	87.5	77.5	95.0
	≥200	13.3	35.8	12.5	22.5	5.0

above 50 years had significantly ($P < 0.05$) high serum triglyceride than all other control groups (Table 7.26).

t-Test was applied for comparison of serum triglyceride level in different study groups with control subjects (Table 7.27). All study groups except females below 50 years of age had significantly high serum triglycerides than control group (Table 7.28).

Reference

Expert Panel on Detection, Evaluation, and Treatment of High Blood Cholesterol in Adults (2001) Executive Summary of the Third Report of the National Cholesterol Education Program (NCEP) Expert Panel on Detection, Evaluation, and Treatment of High Blood Cholesterol in Adults (Adult Treatment Panel III). JAMA 285:2486–2497

Chapter 8
Discussion: Minerals

8.1 Copper

Copper is an essential human micronutrient for enzymes that catalyze oxidation reduction reactions (Linder and Hazegh-Azam 1996). Almost all of the copper is bound to proteins in the body. The total amount of copper in human body is about 75–100 mg, with highest concentrations in the liver and brain tissues. About 90 % of the average 100 μg of copper in the blood is in the form of ceruloplasmin. It is a multi-copper protein widely distributed in vertebrates, whose synthesis and secretion can be markedly increased during inflammation, diabetes, and cardiovascular disorders. It occurs mainly in the plasma and plays an important role in iron homeostasis (Frieden and Hsieh 1976; Gitlin 1998). When ceruloplasmin is low, the mobilization of Fe to transferrin can be impaired, and Fe accumulates in tissues as has been observed in humans lacking functional ceruloplasmin. Elevated copper concentrations may be related to CHD in at least two ways. Oxidation and free radical formation are the two components of atherogenesis. Copper oxidizes LDL cholesterol, increasing its atherogenicity (Heinecke et al. 1984). It was proposed that copper deficiency rather than copper excess is a risk factor for IHD including lipid abnormalities (Klevay 1989). A deficiency of copper can also decrease the activities of certain non-copper-containing enzymes of the oxidant defense system including catalase and selenium-dependent glutathione peroxidase. Descriptive statistics of serum copper in different study groups (Table 6.1) indicated that except control group, in all study groups serum copper level varied directly with age. Analysis of variance (Table 6.2) indicated a significant ($P < 0.05$) effect of age and gender on serum copper of all the study groups except the control. The normal serum level of copper has been reported as 80–120 μg/dL. All the reference values for serum copper derived from the literature are reported from the developed world.

Serum concentration of copper, in the case of control group, was almost half of the lower limit, reported for the normal. Yet the individuals belonging to this age group did not manifest any clinical problem. Females in control groups on an

N.B. Rizvi and S.A. Nagra, *Minerals and Lipids Profiles in Cardiovascular Disorders in South Asia*, DOI 10.1007/978-3-642-34249-3_8,

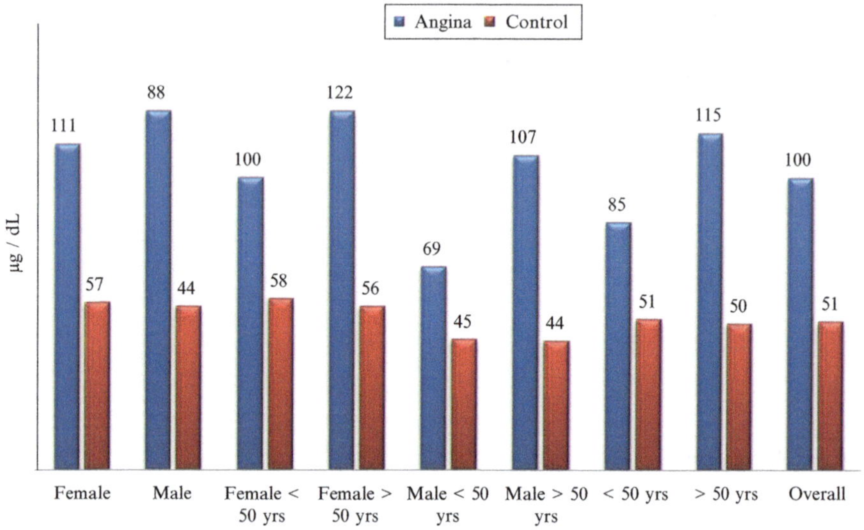

Fig. 8.1 Comparison of mean level of serum copper (μg/dL) in angina patients and control group

average had higher serum copper values than their male counterparts. However, differences were nonsignificant statistically.

Multiple comparisons of mean of serum concentration of copper in angina patients belonging to different age and gender have been given in Table 6.3. A nonsignificant difference was observed in the serum copper level of males above 50 years and females of either age group. All other differences were statistically significant ($P < 0.05$). Almost similar trend was observed in multiple comparisons of mean of serum copper level in angina patients with coexistent DM (Table 6.4).

When compared with the control (Table 6.7), all the study groups had significantly ($P < 0.05$) high serum copper. Graphical comparisons of serum copper in different cardiovascular disorders and control (Figs. 8.1, 8.2, 8.3, 8.4, and 8.5) further substantiate this fact. Variation in serum copper concentration of hypertensive patients as affected by age and gender has been summarized in Fig. 8.5. The effect of age and gender was quite evident on all other CV disorders except control group.

A large body of evidence suggests that copper has essential structural and functional roles throughout the cardiovascular system. These findings are in line with those of Helgeland et al. (1982), who reported that women with CHD had higher copper concentration than the men do (Fischer et al. 1990; Milne and Johnson 1993). An increase in serum copper with age is in agreement with the studies of McMaster et al. (1992) and Campbell (2006). Elevation in serum copper with age had also been studied by Noto et al. (1983) and reported higher serum levels of copper in older subjects. It has been postulated that copper may be linked with the onset and development of atherosclerosis because of its known catalytic function in lipid peroxidation (Kosar et al. 2007). Copper oxidizes LDL cholesterol, increasing its atherogenicity (Heinecke et al. 1984).

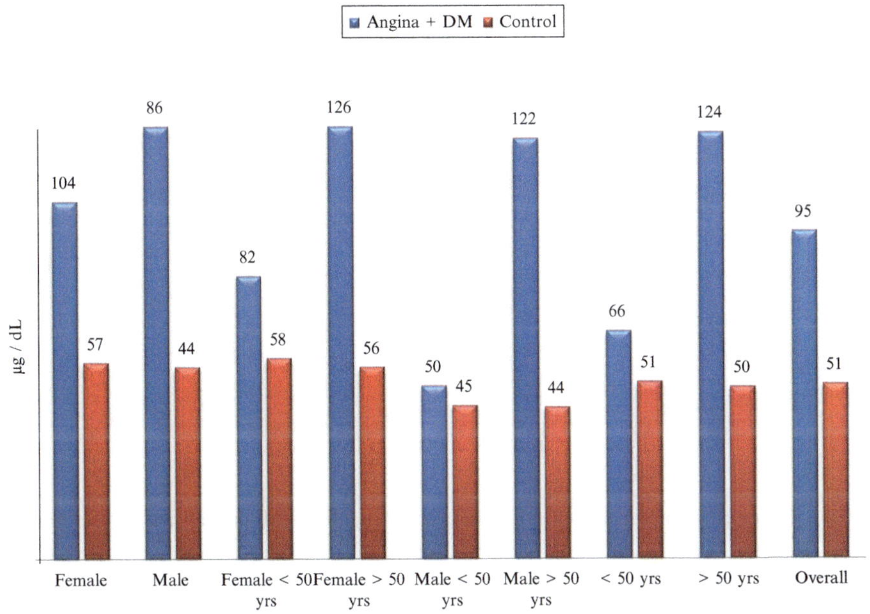

Fig. 8.2 Comparison of mean level of serum copper (µg/dL) in angina patients with coexistent DM and control group

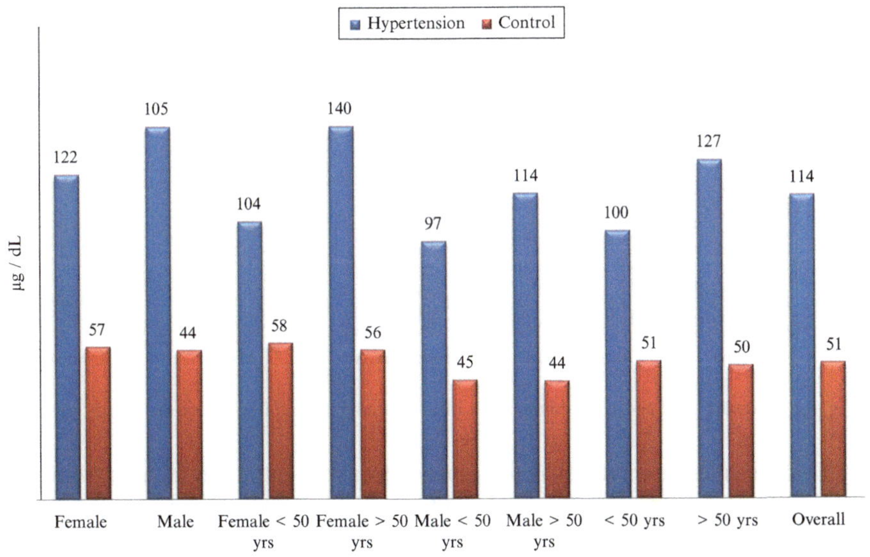

Fig. 8.3 Comparison of mean level of serum copper (µg/dL) in hypertensive patients and control group

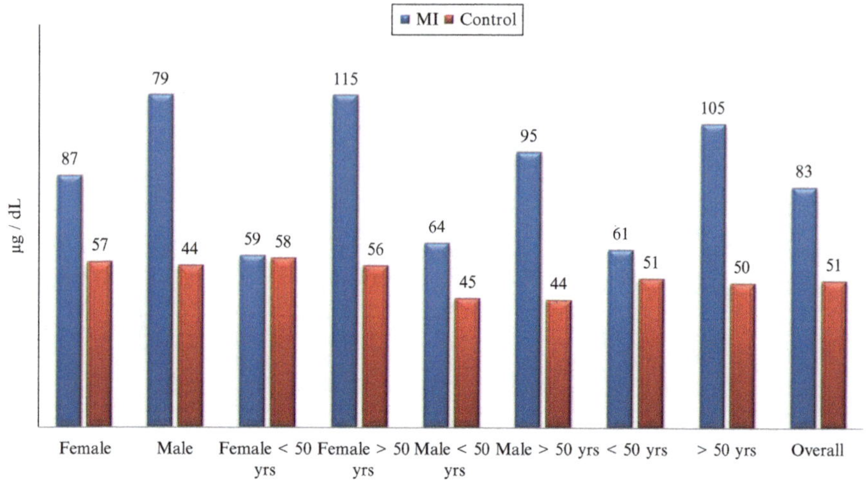

Fig. 8.4 Comparison of mean level of serum copper (μg/dL) in MI patients and control group

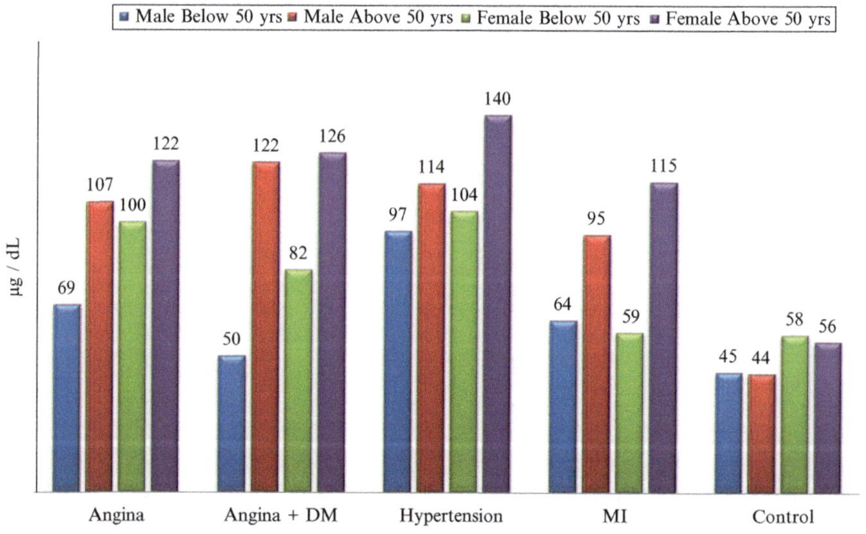

Fig. 8.5 Gender- and age-wise comparison of serum copper (μg/dL) in different study group

Diabetes mellitus and elevated plasma copper concentrations are risk factors for cardiovascular disease (Shukla et al. 2006). In control group, irrespective of gender and age, diminished copper levels were observed (Table 6.1). The results of the study are also in line with those of Tawfeeq et al. (2008) who reported significantly ($P < 0.01$) higher serum copper levels in diabetics as compared to the control group. Similarly, Al-Saleh et al. (2005) and Ekin et al. (2003) have also substantiated that the serum copper was significantly ($P < 0.01$) elevated in diabetic subjects. A plausible explanation has been given by Cooper et al. (2005) who reported that the extracellular superoxide dismutase, an important antioxidant enzyme containing both copper and zinc, was elevated in diabetic subjects, and its activity correlated strongly with the interaction between serum copper and glycosylated hemoglobin. They suggested that CVD complications in diabetes might be better controlled by therapeutic strategies that focus on lowering plasma glucose and loosely bound systemic copper. Diabetes as such has been reported to alter copper status, although differences in trace element levels occurring as a result of diabetes have not been confirmed (Saari 2000). Our results are invariant to those of Fujimoto (1987) who failed to find any definite relationship between copper concentrations and the clinical status of the patients with diabetes mellitus.

Hypertension is one of the leading causes of death and may lead to stroke and heart failure. There may be several associations with effects of copper deficiency on the vascular system with hypertension, such as loss of elasticity and atherogenesis. In some human studies, copper depletion has been found to show abnormalities in blood pressure and electrocardiograms, although this finding is not universal (Institute of medicine 2002). Multiple comparisons of mean serum concentration of copper in hypertensive patients belonging to different age and gender have been given in Table 6.5. Age did not affect the serum copper level in hypertensive male. Below 50 years gender did not appear to affect the copper concentration in serum.

Similarly the serum copper level did not differ significantly in males above 50 years and females less than 50 years. All other comparisons were statistically significant ($P < 0.05$). A high copper level in the tissues has been positively correlated with cardiovascular diseases and hypertension. Comparison of copper level between cases and control showed a significant elevation in serum of all hypertensive (Table 6.7). An increase in serum copper concentration was seen with increasing age in hypertensive patients, especially elderly hypertensive females (139.5 ± 37.5), as compared to normal individuals. Females generally had higher serum copper concentrations than their male counterparts (Milne and Johnson 1993; Fischer et al. 1990). A general increase in serum copper level with advancing age is supported by the work of Uza et al. (1984) who reported a rise in copper levels in the elderly patients with hypertension. Our results are also in line with the study of Akyüz et al. (1993), who reported higher copper levels in the groups of hypertensive patients compared to the control group ($P < 0.001$). They also claimed that copper level goes up in severe heart disease.

Multiple comparisons of mean serum concentration of copper in patients with myocardial infarction belonging to different age and gender have been given in Table 6.6. Below 50 years of age gender did not appear to affect the copper concentration in serum. All other comparisons were statistically significant ($P < 0.05$). Except in males below 50 years and females below 50 years, the effect of age and gender was significantly ($P < 0.05$) conspicuous. Postmenopausal females showed an increase in copper levels (114.9 ± 30.2) as compared to males and age groups when compared with control. The hormonal status, mainly the use of oral contraceptives, can be associated with the increase in serum copper concentration in females. An increase in patient's serum copper levels has been considered a part of a specific defense mechanism to provide more copper at the site of infarction to reduce its size and the extent of damage (Kanabrocki et al. 1964). Tan et al. (1992) reported a significantly higher ($P < 0.001$) copper concentration in patients (138 μg/dL) than in controls (98 μg/dL). Epidemiological studies have reported that elevated serum copper concentrations are an independent risk factor for IHD (Salonen et al. 1991). A decrease in zinc level and an increase in copper levels were observed in acute myocardial infarction patients (Salonen et al. 1991; Vilanova et al. 1997). Several case–control studies concerning copper have found elevated serum copper concentrations in patients with myocardial infarction compared with healthy controls (Kromhout et al. 1985; Singh et al. 1985; Tiber et al. 1986).

8.2 Magnesium

Magnesium (Mg) is the most abundant mineral in the body and is essential for good health. Approximately 50 % of total body magnesium is found in bone. Only 1 % of magnesium is found in blood, but the body works very hard to keep blood levels of magnesium constant (Rude 1998). Most of the serum magnesium is bound to chelators. Roughly 33 % is bound to proteins, and 5–10 % is not bound. The role of magnesium in CVD-related problems is to maintain normal muscle and nerve function and keeps heart rhythm steady. It also helps regulate blood sugar levels, promotes normal blood pressure, and is known to be involved in energy metabolism and protein synthesis (Saris et al. 2000). There is strong evidence that magnesium status is important in the pathogenesis and treatment of cardiovascular disorders and diabetes mellitus (Altura and Altura 1985; Seelig 1980; McCarron 1982).

Descriptive statistics regarding the level of magnesium in control and different study groups has been summarized in Table 6.8. Normal serum level of Mg has been reported as 1.8–3 mg/dL. However, it is evident from Table 6.8 that serum magnesium level in control group was in the lower normal limit reported in literature. In all study groups, on an average, these values were quite lower than control group.

Analysis of variance showed a significant ($P < 0.05$) effect of age groups and gender on serum magnesium level in all the study groups and controls (Table 6.9). Multiple comparisons of magnesium levels in angina patients showed significant difference ($P < 0.05$) in males below 50 and females above 50, males above 50 and females below 50, and males and females above 50. A nonsignificant effect of age was observed in the case of males and in below 50 years of age (Table 6.10). However, comparison of serum magnesium level in different study groups with control (Fig. 8.6; Table 6.15) indicated a significantly low ($P < 0.05$) level of magnesium in angina patients. Elderly females had higher Mg level as compared to other gender and age groups. Our findings agree with those of Altura and Altura (1990) who observed that low concentration of tissue magnesium has been associated with greater CHD occurrence. Decreased magnesium retention has also been reported in patients with variant angina (Goto et al. 1990).

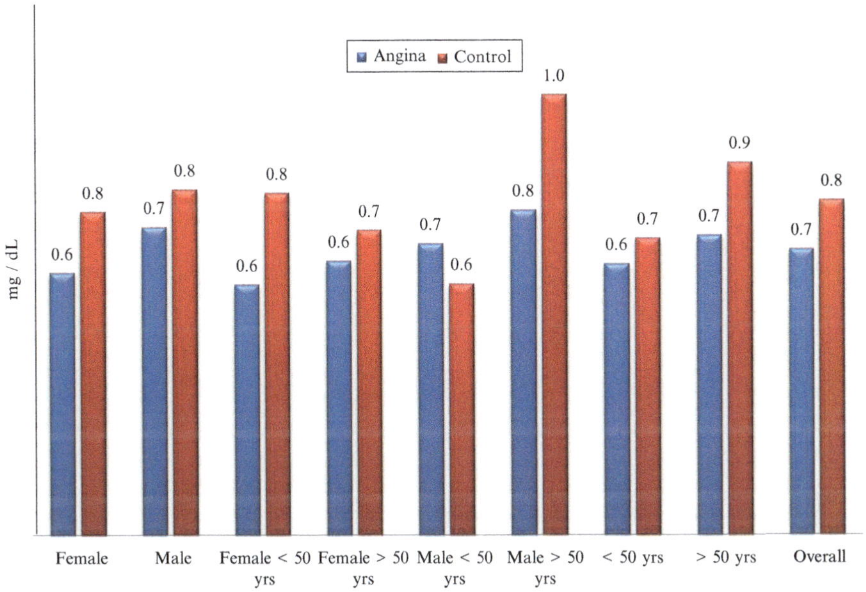

Fig. 8.6 Comparison of mean level of serum magnesium (mg/dL) in angina patients and control group

Multiple comparisons of means of serum magnesium level within angina patients having coexistent diabetes mellitus have been given in Table 6.11. Within any one particular gender effect of age has been observed as nonsignificant. Serum magnesium was significantly low ($P < 0.05$) in all the study groups when compared with the control (Fig. 8.7; Table 6.15). The role of magnesium in carbohydrate metabolism is well established now. It may influence the release and activity of insulin. Patients with diabetes mellitus have been shown to have reduced tissue and serum magnesium concentrations, and this deficiency correlates with difficulty in controlling blood glucose (Nadler et al. 1992). Our results are partially invariant to those of Durlach and Collery (1984) who reported that not all diabetics have both serum and intracellular magnesium deficiencies, and the level differs depending on the type of diabetes and gender of the patient (Levin et al. 1981). Serum magnesium was found to be reduced in diabetic patients (Mather et al. 1979; Schnack et al. 1992; Resnick et al. 1993). Some observational studies have associated higher blood levels of magnesium with lower risk of coronary heart disease (Ford 1999; Liao et al. 1998). Similar trend of variation in serum Mg was observed in multiple comparisons of means of serum magnesium in hypertensive patients (Table 6.12). Serum magnesium was significantly ($P < 0.05$) low in all the hypertensive cases when compared with the control group (Fig. 8.8; Table 6.15). Magnesium may be physiologically important in blood pressure regulation, whereas changes in magnesium levels could contribute to pathological processes underlying hypertension. Postmenopausal females had a significantly ($P < 0.05$) high magnesium levels as

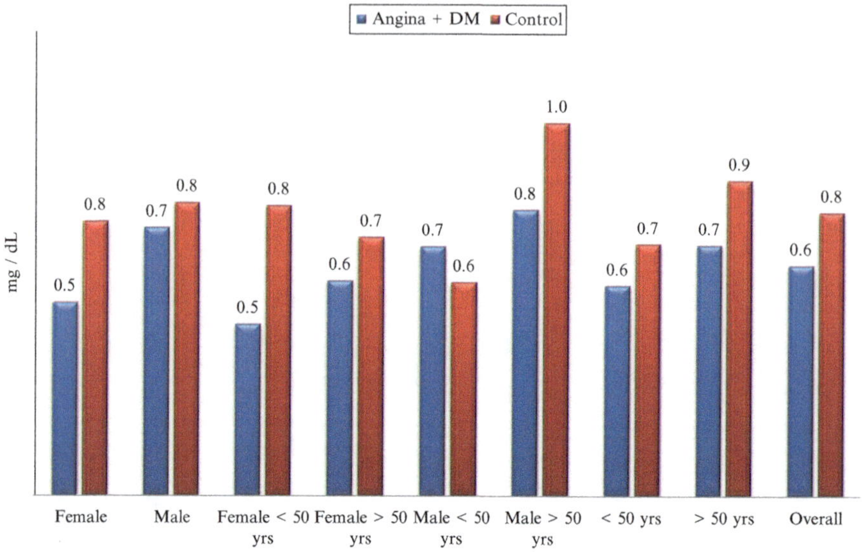

Fig. 8.7 Comparison of mean level of serum magnesium (mg/dL) in angina patients with coexistent DM and control group

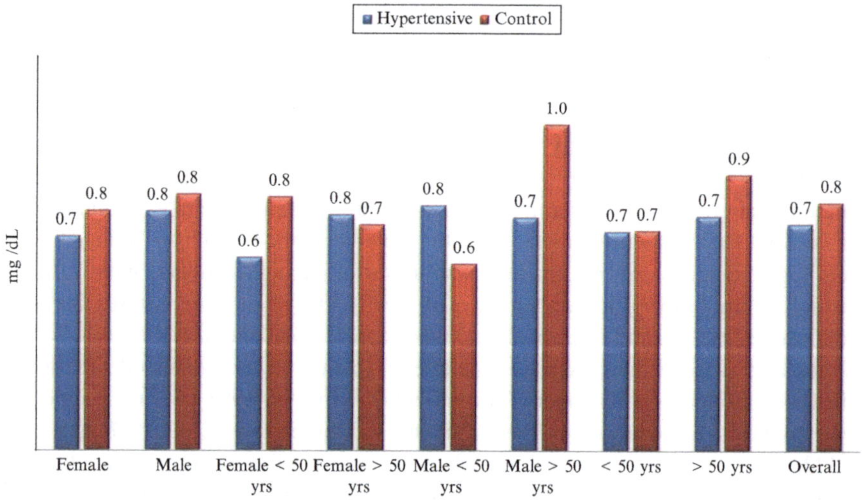

Fig. 8.8 Comparison of mean level of serum magnesium (mg/dL) in hypertensive patients and control group

compared to other age groups and gender (Table 6.12). A large prospective study among nurses in the USA demonstrated that the risk of developing hypertension is reduced by 23 % in women with a magnesium intake of more than 300 mg/day compared with a magnesium intake of less than 200 mg/day (Suter 1999).

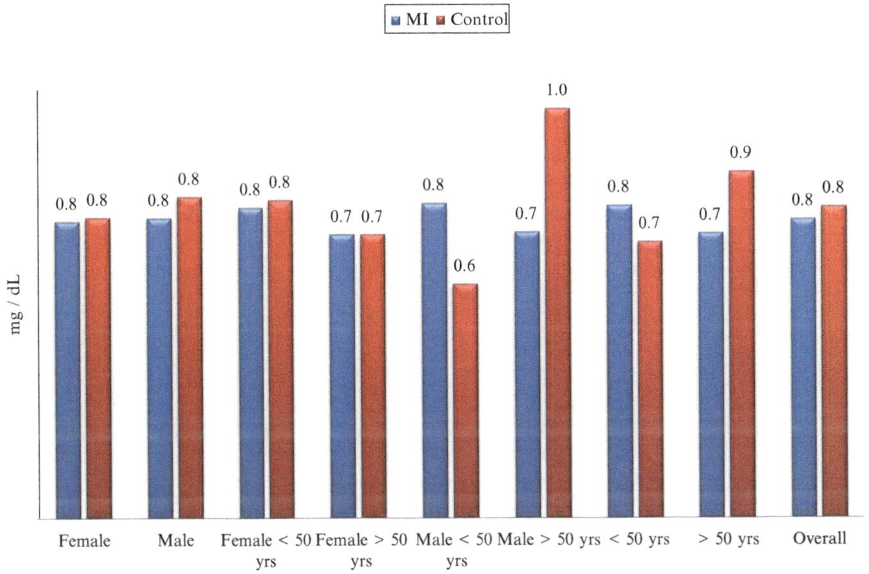

Fig. 8.9 Comparison of mean value of serum magnesium in (mg/dL) MI patients and control group

Epidemiological studies have linked hypertension and hypertensive heart diseases, as well as ischemic heart diseases, with "soft water," low in Mg^{2+}, and protection against cardiovascular disease with "hard water," high in Mg^{2+} (Elwood and Pickering 2002).

It has also been reported that diets rich in magnesium can reduce blood pressure levels, especially in older individuals (Geleinjoise et al. 1996; Ascherio et al. 1996). Our results, however, are partially not in line with those of Cappuccio et al. (1985) and Ferrara et al. (1992), who reported magnesium depletion in hypertension. They observed no differences in serum magnesium levels of normal and hypertensive patients. These differences and variations can be attributed to the various environmental factors and geographical distribution of the data.

Multiple comparisons of mean of serum magnesium level within patients of myocardial infarction have been summarized in Table 6.13. All comparisons were nonsignificant statistically with exception of less than 50 years male and female. However, when compared with the control (Figs. 8.9 and 8.10; Table 6.15), all hypertensive cases had significantly low ($P < 0.05$) serum magnesium level except males and females above 50 years. This effect was particularly prominent in males. Data regarding the use of magnesium in patients with acute myocardial infarction are conflicting. Large decreases in concentrations of total Mg^{2+} in serum during AMI were found to be associated with higher peak concentrations of creatine kinase.

On the other hand, Metwalli et al. (1998) has reported higher serum magnesium levels in acute myocardial infarction patients than that of healthy controls.

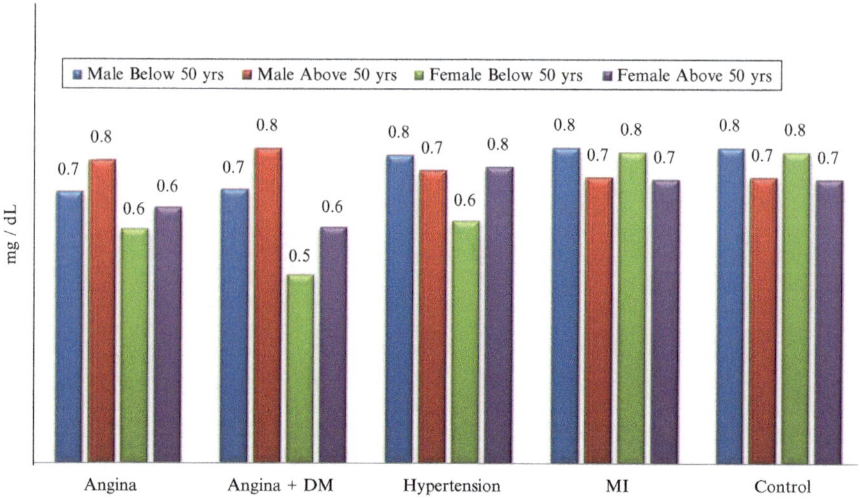

Fig. 8.10 Gender- and age-wise comparison of mean level of serum magnesium (mg/dL) in different study group

Two large cohort studies have reported a significant inverse association between serum magnesium levels and the risk for developing CHD (Ford 1999) or CHD mortality (Liao et al. 1998). In several studies, it was observed that serum total magnesium concentrations were similar between people with various cardiovascular disorders and those who were free of these conditions (Maj-Zurawska 1994). Hypomagnesemia is defined as concentration of Mg^{2+} in serum less than the normal range (McNair et al. 1982). Clinically, hypomagnesemia may be defined as a serum Mg concentration $\leq$1.6 mg/dL or >2 SD below the mean of the general population (Pham et al. 2005). Low levels of magnesium in the blood may mean there is not enough magnesium in the diet, the intestines are not absorbing enough magnesium, or the kidneys are excreting too much magnesium. In general, serum magnesium levels do not adequately reflect a patient's magnesium status and cannot be used to monitor therapeutic interventions (Stühlinger 2002). Average serum Mg level in local population appears to be either lower or very close to the lower normal value. The serum Mg of local population needs to be monitored carefully along with intake so that some dietary interventions may be adopted if indicated.

8.3 Selenium

Selenium was considered highly toxic to animals and humans, initially. However, in 1957 it was recognized as an essential element (U.S. Department of Health and Human Services 2003). Its dietary requirements are very small. It is an essential trace mineral, involved in protection against oxidative damage via selenium-dependent enzyme glutathione peroxidase and other selenoproteins which are important antioxidant enzymes (Rayman 2000). The antioxidant properties of selenoproteins help prevent cellular damage from free radicals. Selenium functions as an antioxidant scavenging H_2O_2 and by reducing lipid hydroperoxides to their subsequent less reactive end products (Rotruck et al. 1973). Selenium intake varies widely around the world mainly because of geographical variations in selenium content of the soil (Rayman 2005; Combs 2001). The recommended dietary allowance for selenium that is estimated to be sufficient to meet the nutritional needs of nearly all healthy adults is 55 μg/day (Rayman 2005). Because of its antioxidant properties, it has long been hypothesized that selenium may prevent cardiovascular disorders and other chronic diseases. Low selenium concentrations may increase risk of cardiovascular disease through various mechanisms, i.e., by shifting prostaglandin synthesis from prostacyclin to thromboxane; low selenium may increase platelet aggregability and vasoconstriction (Rayman 2000; Neve 1996; Huang et al. 2002). Selenium concentration in human serum can be subdivided into three categories: low level below 5–6 μg/dL, intermediate level between 6 and 10 μg/dL, and high level above 10–12 μg/dL (Neve 1991). Descriptive statistics of selenium in different disorders of CVD and study groups have been summarized in Table 6.16. Analysis of variance (Table 6.17) indicated a nonsignificant difference in serum selenium level within the angina, hypertensive, MI patients and control group. However, a significant ($P < 0.05$) difference was observed in serum selenium level within angina patients with coexistent DM. Multiple comparisons of mean selenium in angina patients with coexistent DM (Table 6.18) clearly demonstrated the effect of age and gender.

Multiple comparisons of selenium levels in angina patients with coexistent DM (Table 6.19) clearly demonstrated a significant difference ($P < 0.05$) in selenium levels of males and females below 50, males below 50 and females above 50, males above 50 and females below 50, and females above and below 50, respectively. All other comparisons were nonsignificant.

Comparison of selenium level in control group and patients having various CVD disorders has been summarized in Table 6.20 and Figs. 8.11, 8.12, 8.13, 8.14, and 8.15. None of the study groups showed statistically significant variation with those of the control. It was observed that the mean selenium levels in men having angina with coexistent DM were not significantly different from those found in healthy individuals (Table 6.20). However, in the case of females, variations were observed.

In premenopausal females (58.3 $\pm$ 14.6 μg/dL), mean selenium levels were lower in angina with coexistent DM as compared to postmenopausal females (87.6 $\pm$ 18.9 μg/dL). Our results indicated that all male and female angina patients with coexistent DM had serum selenium levels >12.6 μg/dL (126 μg/L). Serum selenium levels did not differ significantly between angina patients with coexistent DM and healthy individuals (Fig. 8.15; Table 6.20). The results of our study are in line with the work of Holecek et al. (1995), Wang et al. (1995), and Armstrong et al. (1996), who observed serum selenium levels were similar in diabetic patients and control. There is no physiologic reason for concentration differences between the genders, as the status of selenium is exclusively related to the food supply (Swanson

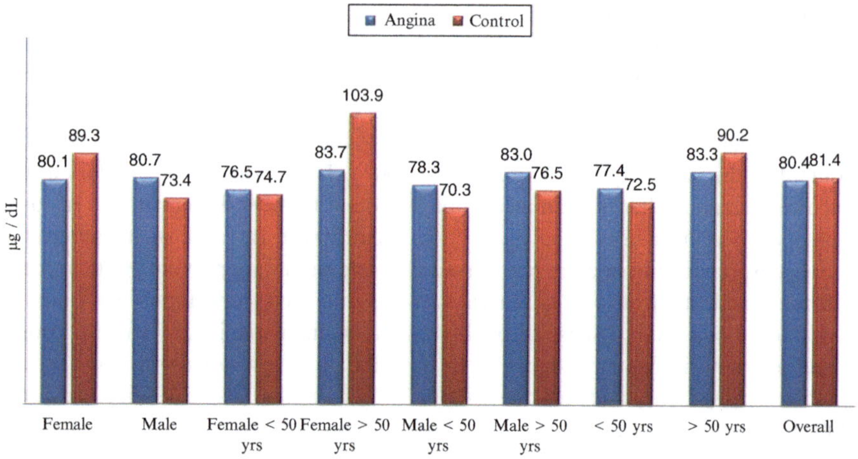

Fig. 8.11 Comparison of mean level of serum selenium (μg/dL) in angina patients and control group

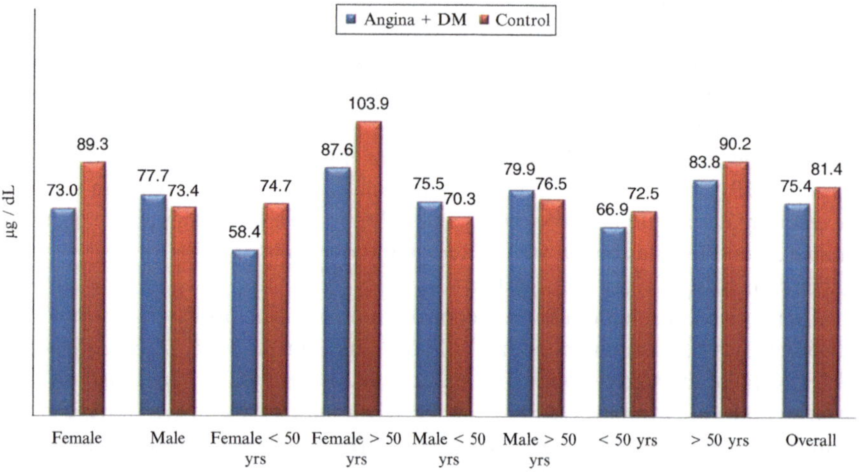

Fig. 8.12 Comparison of mean level of serum selenium (μg/dL) in angina patients with coexistent DM and control group

et al. 1990). Controversial results have been found about the regulation of serum selenium homeostasis in diabetic subjects. Findings of our study indicated that the mean serum selenium levels in men were not significantly different from those found in women. Postmenopausal females, however, showed an increased serum selenium levels as compared to other age groups. In the light of forgoing discussion, it may be concluded that selenium level do not vary with various CVD disorder. Further, mere information of the serum selenium concentration cannot be used as marker/indication for CVD disorder.

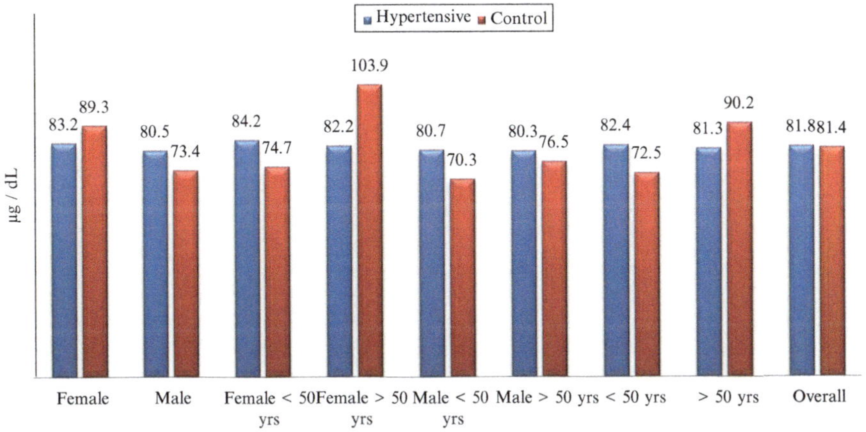

Fig. 8.13 Comparison of mean level of serum selenium (μg/dL) in hypertensive patients and control group

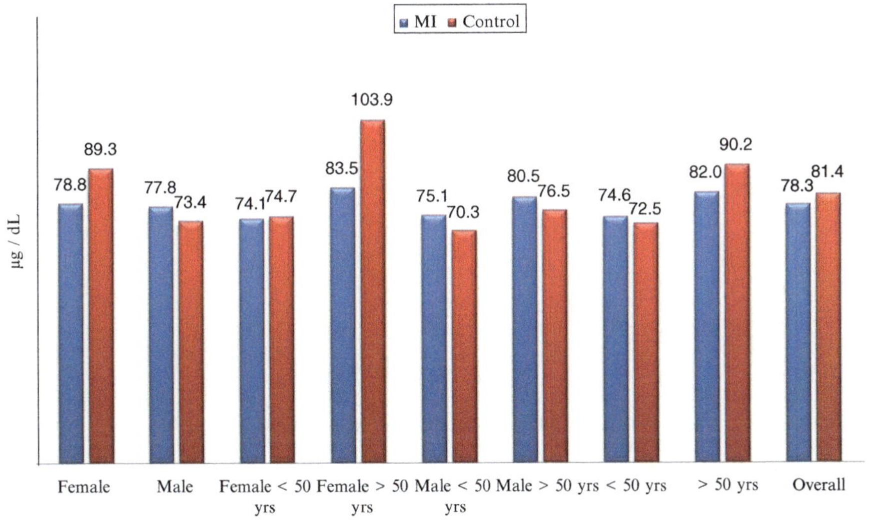

Fig. 8.14 Comparison of mean level of serum selenium (μg/dL) in MI patients and control group

8.4 Zinc

Zinc is an important component of biomembranes and an essential cofactor in a variety of enzymes (Bettger and O'Dell 1981). It also binds and specifically modulates the activity of many membrane receptors, transporters, and channels (Choi 1998). Zinc has antioxidant-like properties. As an antioxidant, zinc has membrane-stabilizing properties and is said to preserve endothelial function (Meerarani et al. 2000; Hennig et al. 1999). Zinc is among the most abundant trace elements in mammals. It is generally considered that adequate zinc helps to

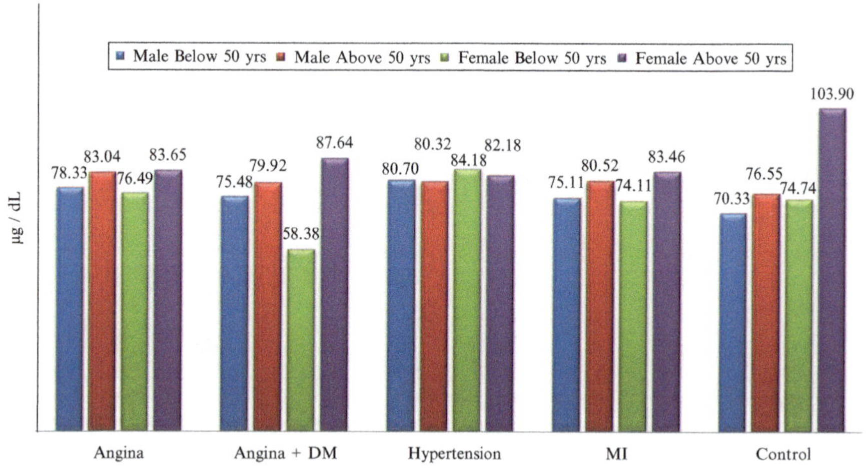

Fig. 8.15 Gender- and age-wise comparison of serum selenium (µg / dL) in different study groups

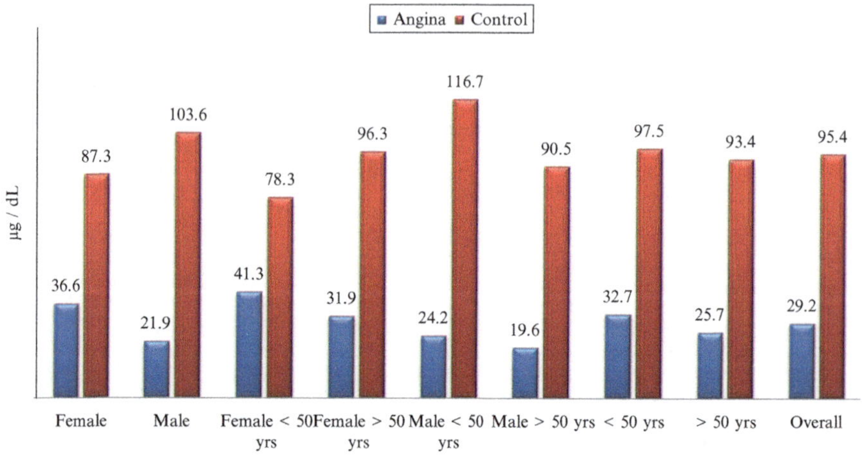

Fig. 8.16 Comparison of mean level of serum zinc (µg/dL) in angina patients and control group

keep the artery walls flexible, and deficiency of zinc results in hardening of arteries. Low zinc levels also allow the tissue sodium level to rise, which can contribute to high blood pressure and fluid retention. Superoxide dismutase, an enzyme containing both copper and zinc, is found in almost all oxygen-utilizing cells and is essential for catalyzing reactions for removing the highly reactive superoxide anion (Ma and Betts 2000). Hennig et al. (1996) reported that zinc has antiatherogenic properties by preventing metabolic physiologic derangements of the vascular endothelium. Because of its antioxidant and membrane-stabilizing properties, zinc appears to be crucial for the protection against cell-destabilizing agents such as polyunsaturated lipids and inflammatory cytokines.

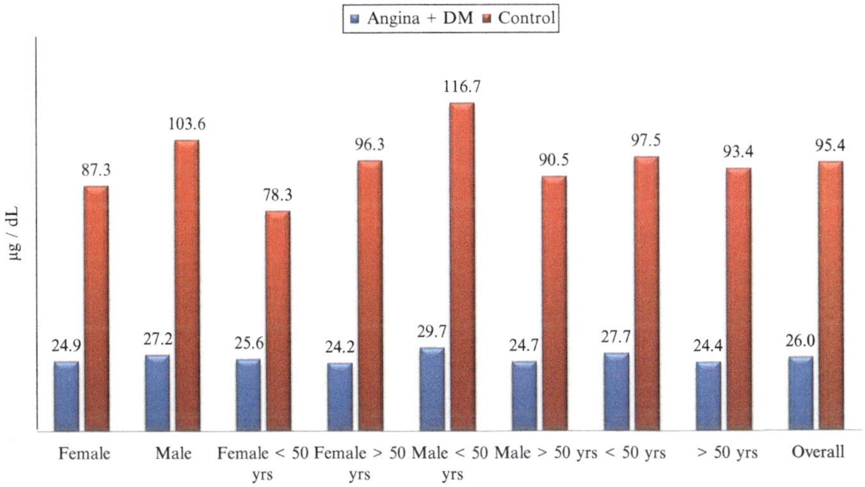

Fig. 8.17 Comparison of mean level of serum zinc (µg/dL) in angina patients with coexistent DM and control group

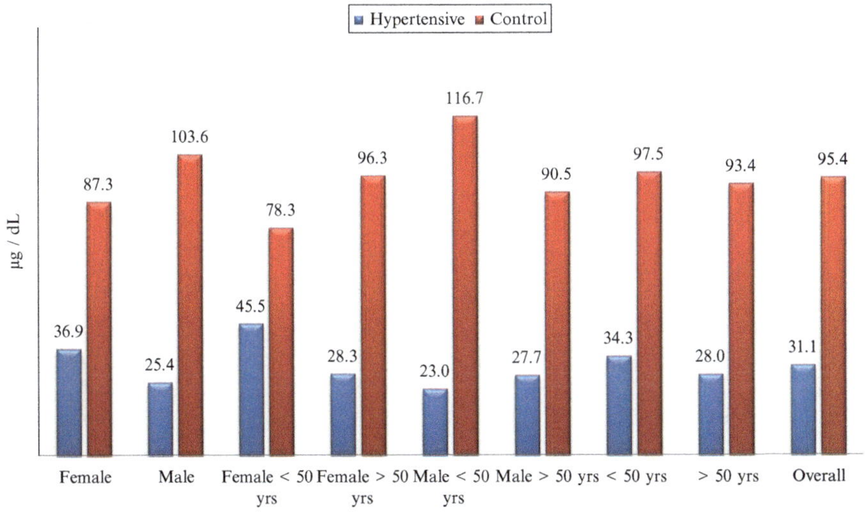

Fig. 8.18 Comparison of mean level of serum zinc (µg/dL) in hypertensive patients and control group

Descriptive statistics of serum zinc in various cardiovascular disorders, age groups, and gender has been summarized in Table 6.21. The analysis of variance for serum zinc level (Table 6.22) indicated a significant ($P < 0.05$) effect on age and gender within angina and hypertensive patients. Within other cardiovascular disorders and control group, a nonsignificant effect of age and gender was observed. However, when compared with the control, patients in all categories of

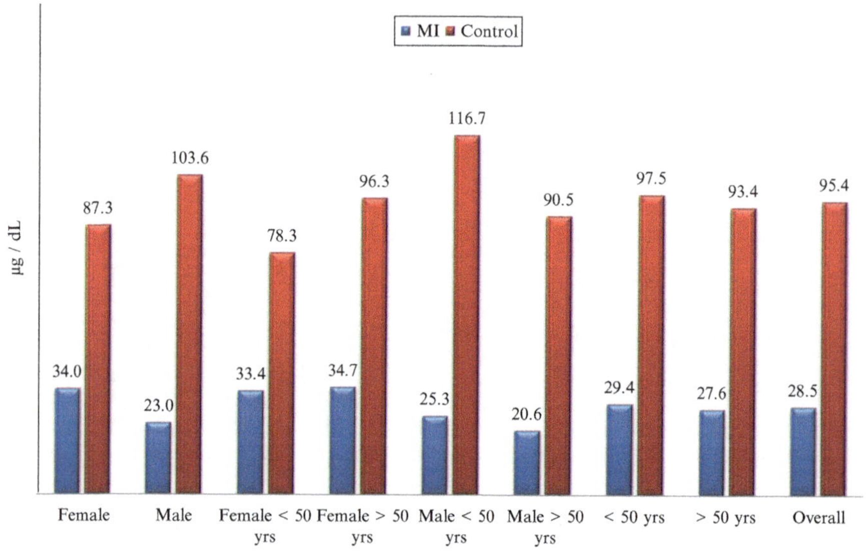

Fig. 8.19 Comparison of mean level of serum zinc (μg/dL) in MI patients and control group

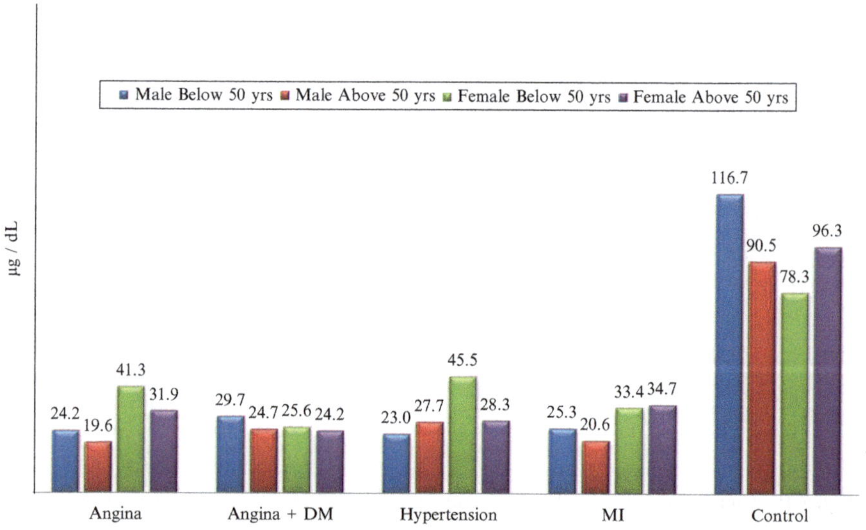

Fig. 8.20 Gender- and age-wise comparison of mean level of serum zinc (μg/dL) in different study group

cardiovascular disorders had significantly low serum zinc (Table 6.25 and Figs. 8.16, 8.17, 8.18, 8.19, and 8.20). Multiple comparisons of mean serum concentration of zinc in angina patients belonging to different age and gender have been summarized in Table 6.24. Males below 50 years of age had significantly

($P < 0.05$) lower serum zinc than females of the same age group. Males above 50 years of age had significantly ($P < 0.05$) lower serum zinc than females of either age.

These findings are in agreement with the work of Kazemi-Bajestani et al. (2007) who observed a consistent change in serum zinc concentration with age in patients with cardiovascular disorders. A fall in plasma zinc with age has been reported and attributed to a decline in the rate of absorption or an accelerated clearance from the plasma (Bales et al. 1986). Zinc may play an antiatherogenic role through the inhibition of oxidative stress and apoptosis of endothelial cells during inflammatory conditions (Hennig et al. 1999; Meerarani et al. 2000). A few cross-sectional studies in patients with various forms of CVD have reported a decreased serum zinc concentration (Martin-Lagos et al. 1997; Singh et al. 1998).

Klevay (1975) reported that zinc deficiency may predispose to CHD, and an imbalance in zinc/copper metabolism results in CAD. According to their study, dietary copper deficiency alone, or in association with a high zinc intake, may lead to atherosclerosis.

Zinc has important role in modulating the immune system, and its dysfunction in diabetes mellitus may be related in part to the status of zinc (Mocchegianai et al. 1989). Zinc metabolism seems to be altered in diabetic patients. Comparison of zinc levels between angina patients with coexistent DM and control also showed significantly ($P < 0.05$) low levels of zinc (Table 6.25; Fig. 8.17). On an average, there was no consistent change in serum zinc levels with age in angina patients with coexistent DM (Table 6.21). In normal subjects, a change in serum zinc was observed with age, especially in males below 50 years. Lower concentrations of serum zinc were found to be associated with a higher prevalence of CAD, diabetes, and glucose intolerance in both men and women (Singh et al. 1998). Kinlaw et al. (1983) reported that 25 % of patients with diabetes had zinc deficiency; however, the difference in fasting serum zinc concentrations between diabetic and control subjects (84.2 ± 17.7 vs. 96.4 ± 8.0 μg/dL, respectively) was not significant. Zinc levels were also observed to be significantly lower ($P < 0.05$) in patients with diabetes compared to nondiabetic patients who suffered a heart attack (58.6 ± 6.9 vs. 68.2 ± 6.2 μg/dL). The clinical significance and evaluation of zinc in diabetes mellitus remains controversial. Marjani (2005) reported a decrease in zinc levels ($P < 0.05$) in type II DM patients when compared with control. Some studies described decreased plasma Zn in type II diabetes mellitus patients (Walter et al. 1991; Evliaoglu et al. 2002), whereas others showed no significant differences (Zargar et al. 1998). Alterations of zinc metabolism result in reduced availability of zinc that might be expected to contribute to tissue damage observed in diabetes (Sumovski et al. 1989). Diabetes is usually associated with decreased serum zinc concentration (Levine et al. 1983; Canfield et al. 1984; Aguilar et al. 2007; Kruse-Jarres and Rükgauer 2000).

Multiple comparisons of mean serum concentration of zinc in hypertensive patients belonging to different age and gender have been given in Table 6.24. Effect of age was nonsignificant in the case of males. However, females above 50 years of age had significantly ($P < 0.05$) high serum zinc than below 50 years of age. Hypertensive patients had significantly ($P < 0.05$) low zinc level when compared with control (Table 6.25, Fig. 8.18). It is reported that the diastolic and systolic blood pressure of both men and women were inversely related with serum zinc (Harlan 1988). The risk of copper deficiency is also thought to be increased by the consumption of excess zinc, as these two elements can be biologically antagonist (Hurley et al. 1983).

Rubio-Luengo et al. (1995) observed a decrease in serum Zn concentration in hypertensive patients. A nonsignificant difference was observed in zinc level of various study groups suffering from MI. However, when compared with control (Table 6.25; Fig. 8.19), MI patients had significantly ($P < 0.05$) low Zn levels in both the gender and age groups. These findings are in line with those of Handjani et al. (1974) who reported a fall in serum zinc after acute tissue injury including myocardial infarction. The precise reason of low serum zinc levels in myocardial infarction is still unknown. Halsted and Smith (1970) reported that serum zinc levels decrease in MI patients within 24–48 h, and this decrease persists for 2 weeks. Kazi et al. (2008) reported an increase in serum copper and a decrease in zinc concentration in MI patients. It was reported that decrease in serum zinc level occur within first three postinfarction days, which rose back to near normal by the tenth day (Jain and Mohan 1991). We also observed that serum copper and zinc vary in an inverse relation to each other. Our results, however, do not agree with those of Metwalli et al. (1998) who reported a nonsignificant difference in serum zinc levels in myocardial patients when compared with control. In the light of foregoing results and discussion, it is evident that serum copper level is significantly elevated ($P < 0.05$) in all groups of cardiovascular patients when compared with the control. Females above 50 years of age had the highest serum copper level followed by their male counterparts. On the contrary, serum zinc was significantly low in all categories of cardiovascular patients when compared with the control. It appears as both the trace minerals vary inversely in cardiovascular disorder. The correlation of the mutual trend of variation of the two minerals may open up the new vistas of research in the future. Serum magnesium was close to the lower normal limit in control as well as study groups and did not differ strikingly when compared with the control. However, females above 50 years in control group had the highest ($P < 0.05$) serum selenium level. In our study, magnesium and selenium did not appear to be associated much with cardiovascular problems.

References

Aguilar MV, Saavedra P, Arrieta FJ, Mateos CJ, González MJ, Meseguer I (2007) Plasma mineral content in type-2 diabetic patients and their association with the metabolic syndrome. Ann Nutr Metab 51(5):402–406

Akyüz F, Önder E, Erden M (1993) Evaluation of serum magnesium, zinc, copper and ascorbic acid levels in patients with hypertension and atherosclerotic heart diseases. Turk J Med Res 11 (6):273–276

Al-Saleh E, Nandakumaran M, Al-Shammari M, Makhseed M, Sadan T, Harouny A (2005) Maternal-fetal status of copper, iron, molybdenum, selenium and zinc in insulin-dependent diabetic pregnancies. Arch Gynecol Obstet 271(3):212–217

Altura BM, Altura BT (1985) New perspectives on the role of magnesium in the pathophysiology of the cardiovascular system. Magnesium 4:226–244

Altura BM, Altura BT (1990) Magnesium and the cardiovascular system: experimental and clinical aspects updated. In: Sigel H, Sigel A (eds) Compendium on magnesium and its role in biology, nutrition, and physiology. Marcel Dekker, New York, pp 359–416

Armstrong AM, Chestnutt JE, Gormley MJ, Young IS (1996) The effect of dietary treatment on lipid per oxidation and antioxidant status in newly diagnosed non-insulin dependent diabetes. Free Radic Biol Med 21:719–726

Ascherio A, Hennekens C, Willett WC, Sacks F, Rosner B, Manson J, Witteman J, Stampfer MJ (1996) Prospective study of nutritional factors, blood pressure and hypertension among US women. Hypertension 27(5):1065–1072

Bales CW, Steinman LC, Freeland-Graves JH, Stone JM, Young RK (1986) The effect of age on plasma zinc uptake and taste acuity. Am J Clin Nutr 44:664–669

Bettger W, O'Dell B (1981) A critical physiological role of zinc in the structure and function of biomembranes. Life Sci 28:1425–1438

Campbell A (2006) The role of aluminum and copper on neuro-inflammation and Alzheimer's disease. J Alzheimers Dis 10:165–172

Canfield WK, Hambidge KM, Johnson LK (1984) Zinc nutriture in type I diabetes mellitus: relationship to growth measures and metabolic control. J Pediatr Gastroenterol Nutr 3(4):577–584

Cappuccio FP, Markandu ND, Beynon GW, Shore AC, Sampson B, MacGregor GA (1985) Lack of effect of oral magnesium on high blood pressure: a double blind study. Br Med J 291:235–238

Choi DW (1998) Zinc and brain injury. Ann Rev Neurosci 21:347–375

Combs GF Jr (2001) Impact of selenium and cancer-prevention findings on the nutrition-health paradigm. Nutr Cancer 40:6–11

Cooper GJ, Chan YK, Dissanayake AM, Leahy FE, Keogh GF, Frampton CM, Gamble GD, Brunton DH, Baker JR, Poppitt SD (2005) Demonstration of a hyperglycemia-driven pathogenic abnormality of Copper homeostasis in diabetes and its reversibility by selective chelation: quantitative comparisons between the biology of copper and eight other nutritionally essential elements in normal and diabetic individuals. Diabetes 54(5):1468–1476

Durlach J, Collery P (1984) Magnesium and potassium in diabetes and carbohydrate metabolism. Magnesium 3:315–323

Ekin S, Mert N, Gunduz H, Meral I (2003) Serum sialic acid levels and selected mineral status in patients with type 2 diabetes mellitus. Biol Trace Elem Res 94(3):193–201

Elwood PC, Pickering J (2002) Magnesium and cardiovascular disease: a review of epidemiological evidence. J Clin Basic Cardiol 5:61–66

Evliaoglu O, Kilicaslan N, Uzuncan N, Karaca B, Kocaclebi A, Yensel N, Inci S (2002) Serum levels of Cu, Zinc, Mg in type 1and 2 diabetic patients. In: 17th Turkish National Biochemical Congress, pp 285–286

Ferrara LA, Iannuzzi R, Castaldo A, Iannuzzi A, Dello Russo A, Mancini M (1992) Long-term magnesium supplementation in essential hypertension. Cardiology 81:25–33

Fischer PWF, L'Abbé MR, Giroux A (1990) Effects of age, smoking, drinking, exercise and estrogen use on indices of copper status in healthy adults. Nutr Res 10:1081–1090

Ford ES (1999) Serum magnesium and ischemic heart disease: findings from a national sample of US adults. Int J Epidemiol 28:645–651

Frieden E, Hsieh HS (1976) Ceruloplasmin: the copper transport protein with essential oxidase activity. Adv Enzymol Relat Areas Mol Biol 44:187–236

Fujimoto S (1987) Studies on the relationship between blood trace metal concentration and the clinical status of patients with cerebrovascular disease, gastric cancer and diabetes mellitus. Hokoido Igaku Zasshi 62:913–932

Geleinjoise JM, Witteman JC, den Breeijen JH, Hofman A, de Jong PT, Pols HA, Grobbee DE (1996) Dietary electrolyte intake and blood pressure in older subjects: the Rotterdam Study. J Hypertens 14(6):737–741
Gitlin JD (1998) Aceruloplasminemia. Pediatr Res 44:271–276
Goto K, Yasue H, Okumura K, Matsuyama K, Kugiyama K, Miyagi H, Higashi T (1990) Magnesium deficiency detected by intravenous loading test in variant angina pectoris. Am J Cardiol 65:709–712
Halsted JA, Smith JC (1970) Plasma-zinc in health and disease. Lancet 1(7642):322–324
Handjani AM, Smith JC Jr, Herrmann JB, Halsted JA (1974) Serum zinc concentration in acute myocardial infarction. Chest 65:185
Harlan WR (1988) The relationship of blood lead levels to blood pressure in the US population. Environ Health Perspect 78:9–13
Heinecke JW, Rosen H, Chait A (1984) Iron and copper promote modification of low density lipoprotein by human arterial smooth muscle cells in culture. J Clin Invest 74:1890–1894
Helgeland K, Haider T, Johnsan J (1982) copper and zinc in human serum in Norway. Relationship to geography, sex and age. Scand J Clin Lab Invest 42(1):35–39
Hennig B, Toborek M, McClain CJ (1996) Antiatherogenic properties of zinc: implications in endothelial cell metabolism. Nutrition 12(10):711–717
Hennig B, Meerarani P, Toborek M, McClain CJ (1999) Antioxidant-like properties of zinc in activated endothelial cells. J Am Coll Nutr 18:152–158
Holecek V, Racek J, Jerabek Z (1995) Administration of multivitamin combinations and trace elements in diabetes. Cas Lek Cesk 134:80–83
Huang KC, Lin WY, Lee LT (2002) Four anthropometric indices and cardiovascular risk factors in Taiwan. Int J Obes Relat Metab Disord 26:1060–1068
Hurley LS, Keen CL, Lonnerdal B (1983) Aspects of trace element interactions during development. Fed Proc 42:1735–1739
Institute of Medicine (2002) Dietary reference intakes: vitamin A, vitamin K, arsenic, boron, chromium, copper, iodine, iron, manganese, molybdenum, nickel, silicon, vanadium, and zinc. Food and Nutrition Board. National Academy Press, Washington, DC, pp 224–257
Jain VK, Mohan G (1991) Serum zinc and copper in myocardial infarction. Biol Trace Elem Res 31:317
Kanabrocki EL, Fields T, Decker CF (1964) Neutron activation studies of biological studies: manganese and copper. Int J Appl Radiat 15:175–190
Kazemi-Bajestani SM, Ghayour-Mobarhan M, Ebrahimi M, Moohebati M, Esmaeili HA, Parizadeh MR, Aghacizadeh R, Ferns GA (2007) Serum copper and zinc concentrations are lower in Iranian patients with angiographically defined coronary artery disease than in subjects with a normal angiogram. J Trace Elem Med Biol 21:22–28
Kazi TG, Afridi HI, Kazi N, Jamali MK, Arain MB, Sarfraz RA, Jalbani N, Ansari R, Shah AQ, Memon AU, Khandhro GA (2008) Distribution of zinc, copper and iron in biological samples of Pakistani myocardial infarction (1st, 2nd and 3rd heart attack) patients and controls. Clin Chim Acta 389(1–2):114–119
Kinlaw WB, Levine AS, Morley JE, Silvis SE, McClain CJ (1983) Abnormal zinc metabolism in type II diabetes mellitus. Am J Med 75:273–277
Klevay LM (1975) Coronary heart disease: the zinc/copper hypothesis. Am J Clin Nutr 28:764–774
Klevay LM (1989) Ischemic heart disease as copper deficiency. Adv Exp Med Biol 258:197–208
Kosar F, Taskapan C, Kucukbay Z (2007) Serum levels of selenium, zinc and copper in patients with coronary artery ectasia. Indian Heart J 59(1):38–41
Kromhout D, Wibowo AAE, Herber RFM (1985) Trace metals and coronary heart disease risk indicators in 152 elderly men (The Zutphen Study). Am J Epidemiol 122:378–385
Kruse-Jarres JD, Rükgauer M (2000) Trace elements in diabetes mellitus. Peculiarities and clinical validity of determinations in blood cells. J Trace Elem Med Biol 14(1):21–27

Levin GE, Mather HM, Pilkington TRE (1981) Tissue magnesium status in diabetes mellitus. Diabetologia 21:131–134
Levine AS, McClain CJ, Handwerger BS, Brown DM, Morley JE (1983) Tissue zinc status of genetically diabetic and streptozotocin-induced diabetic mice. Am J Clin Nutr 37(3):382–386
Liao F, Folsom A, Brancati F (1998) Is low magnesium concentration a risk factor for coronary heart disease? The Atherosclerosis Risk in Communities (ARIC) Study. Am Heart J 136:480–490
Linder MC, Hazegh-Azam M (1996) Copper biochemistry and molecular biology. Am J Clin Nutr 63:797S–811S
Ma J, Betts NM (2000) Zinc and copper intakes and their major food sources for older adults in the 1994-96 continuing survey of food intakes by individuals (CSF 11). J Nutr 130:2838–2843
Maj-Zurawska M (1994) Clinical findings on human blood with the KONE ISE for Mg. Scand J Clin Lab Invest 54(suppl 217):69–76
Marjani A (2005) Plasma lipid peroxidation zinc and erythrocyte Cu-Zn super oxide dismutase enzyme activity in patients with type II diabetes mellitus in Gorgan City (South East of the Caspian Sea). Internet J Endocrinol 2(1):1540–2606
Martin-Lagos F, Navarro-Alarcon M, Terres-Martos C, Lopez G, de la Serena H, Lopez-Martinez MC (1997) Serum copper and zinc concentrations in serum from patients with cancer and cardiovascular disease. Sci Total Environ 204:27–35
Mather HM, Nisbet JA, Burton GH (1979) Hypomagnesaemia in diabetes. Chim Clin Acta 95:235–242
McCarron DA (1982) Calcium, magnesium, and phosphorus balance in human and experimental hypertension. Hypertension 4(Suppl III):III-27–III-33
McMaster D, McCrum E, Patterson CC, Kerr MM, O'Reilly D, Evans AE (1992) Serum copper and zinc in random samples of the population of Northern Ireland. Am J Clin Nutr 56:440–446
McNair P, Christensen MS, Christiansen C, Madsbad S, Transbol I (1982) Renal hypomagnesaemia in human diabetes mellitus: its relation to glucose homeostasis. Eur J Clin Invest 12:81–85
Meerarani P, Ramadass P, Toborek M, Bauer HC, Bauer H, Hennig B (2000) Zinc protects against apoptosis of endothelial cells induced by linoleic acid and tumor necrosis factor α. Am J Clin Nutr 71:81–87
Metwalli O, Al-okbi S, Motawi T, El-ahmady O, Abdul-Hafeez S, El-said E (1998) Study of serum metals and lipid profile in patients with acute myocardial infarction. J Islam Acad Sci 11 (1):5–12
Milne DB, Johnson PE (1993) Assessment of copper status: effect of age and gender on reference ranges in healthy adults. Clin Chem 39:883–887
Mocchegianai E, Boemi M, Fumelli P, Fabris N (1989) Zinc-dependent low thymic hormone level in type 1 diabetes. Diabetes 38:932–937
Nadler JL, Malayan S, Loung H, Shaw S, Natarajan RD, Rude RK (1992) Intracellular free magnesium deficiency plays a key role in increased platelet reactivity in type II diabetes mellitus. Diabetes Care 15(7):835–841
Neve J (1991) Methods in determination of selenium states. J Trace Elem Electr Health Dis 5:1–17
Neve J (1996) Selenium as a risk factor for cardiovascular diseases. J Cardiovasc Risk 3:42–47
Noto R, Alicata R, Sfoglian L, Neri S, Bifarella M (1983) A study of cupermia in a group of elderly diabetics. Acta Diabetol Lat 20:81–85
Pham PC, Pham PM, Pham PA, Pham SV, Pham HV, Miller JM, Yanagawa N, Pham PT (2005) Lower serum magnesium levels are associated with more rapid decline of renal function in patients with diabetes mellitus type 2. Clin Nephrol 63:429–436
Rayman MP (2000) The importance of selenium to human health. Lancet 356:233–241
Rayman MP (2005) Selenium in cancer prevention: a review of the evidence and mechanism of action. Proc Nutr Soc 64:527–542
Resnick LM, Altura BT, Gupta RK (1993) Intracellular and extra cellular magnesium depletion in type II (non insulin-dependent) diabetes mellitus. Diabetologia 36:767–770

Rotruck JT, Pope AL, Ganther HE, Swanson AB, Hafeman DG, Hoekstra WG (1973) Selenium: biochemical role as component of glutathione peroxidase. Science 179:538–544

Rubio-Luengo AM, Maldonado-Martin A, Gil Extremera B, Gonzalez-Gomez L, Luna del Castillo JD (1995) Variations in magnesium and zinc in hypertensive patients receiving different treatments. Am J Hypertens 8:689–695

Rude RK (1998) Magnesium deficiency: a cause of heterogeneous disease in humans. J Bone Miner Res 13:749–758

Saari JT (2000) Copper deficiency and cardiovascular disease: role of per oxidation, glycation, and nitration. Can J Physiol Pharmacol 78(10):848–855

Salonen JT, Salonen R, Korpela H, Suntioinen S, Tuomilehto J (1991) Serum copper and the risk of acute myocardial infarction: a prospective population study in men in eastern Finland. Am J Epidemiol 134:268–276

Saris NE, Mervaala E, Karppanen H, Khawaja JA, Lewenstam A (2000) Magnesium: an update on physiological, clinical, and analytical aspects. Clin Chim Acta 294:1–26

Schnack CH, Bauer I, Pregnant P (1992) Hypomagnesaemia in type II (non-insulin dependent) diabetes mellitus is not corrected by improvement of long-term metabolic control. Diabetologia 35:77–79

Seelig MS (1980) Magnesium deficiency in the pathogenesis of disease: early roots of cardiovascular skeletal and renal abnormalities. Plenum, New York, pp 141–266

Shukla N, Maher J, Masters J, Angelini GD, Jeremy JY (2006) Does oxidative stress change ceruloplasmin from a protective to a vasculopathic factor. Atherosclerosis 187(2):238–250

Singh MM, Singh R, Khare A, Gupta MC, Patney NL, Jain VK, Goyal SP, Prakash V, Pandey DN (1985) Serum copper in myocardial infarction-diagnostic and prognostic significance. Angiology 36:504–510

Singh RB, Niaz MA, Rastogi SS, Bajaj S, Gaoli Z, Shoumin Z (1998) Current zinc intake and risk of diabetes and coronary artery disease and factors associated with insulin resistance in rural and urban populations of North India. J Am Coll Nutr 17(6):564–570

Stühlinger HG (2002) Focus on magnesium: magnesium deficiency and cardiovascular disease. Magnesium in cardiovascular disease. J Clin Basic Cardiol 5:55

Sumovski W, Baquerizo H, Rabinovich A (1989) Oxygen free radical scavenger protects rat islet cells from damage by cytokines. Diabetelogica 32:792–796

Suter PM (1999) The effects of potassium, magnesium, calcium and fiber on risk of stroke. Nutr Rev 57(30):84–88

Swanson CA, Longnecker MP, Veillon C, Howe M, Levander OA, Taylor PR (1990) Selenium intake, age, gender, and smoking in relation to indices of selenium status of adults residing in a seleniferous area. Am J Clin Nutr 52:858–862

Tan IK, Chua KS, Toh AK (1992) Serum magnesium, copper, and zinc concentrations in acute myocardial infarction. J Clin Lab Anal 6(5):324–328

Tawfeeq FR, Abbas MR, Abdul Kareem Y (2008) Serum copper, zinc and Cu/Zn ratio in diabetics. Iraqi J Commun Med 21(1):64–68

Tiber AM, Sakhaii M, Joffe CD, Ratnaparkhi MV (1986) Relative values of plasma copper, zinc, lipids and lipoproteins as markers for coronary artery disease. Atherosclerosis 62:105–110

U.S. Department of Health and Human Services (2003) Toxicological profile for selenium. U.S. Department of Health and Human Services, Public Health Service, Atlanta, GA

Uza G, Pavel O, Kovacis A (1984) Serum concentration of Na, K, Ca, Mg, P, Zn and 2u in patients with essential arterial hypertension. Clin Exp Hypertens A 6(8):1415–1429

Vilanova A, Gutiérrez C, Serrat N, Raga X, Paternain JL (1997) Metallothionein, zinc and copper levels: relationship with acute myocardial infarction. Clin Biochem 30(3):235–238

Walter RM, Uriu-Hare JY, Olin KL (1991) Copper, zinc, magnesium status and complications of diabetes mellitus. Diabetes Care 14:1050–1056

Wang WC, Makela AL, Nanto V, Makela P (1995) Serum selenium levels in diabetic children. A follow-up study during selenium-enriched agricultural fertilization in Finland. Biol Trace Elem Res 47:355–364

Zargar AH, Shah NA, Massodi SR (1998) Copper, zinc and magnesium levels in non insulin dependent diabetes mellitus. Postgard Med J 74:665–668

Chapter 9
Discussion: Lipid Profile

9.1 Angina

The link between lipid abnormalities and CVD is well established. A lipid profile measures total cholesterol, HDL cholesterol, LDL cholesterol, and triglycerides (TG). Abnormal blood lipid levels, which are high total cholesterol, high levels of triglycerides, high levels of LDL, and low levels of HDL cholesterol, increase the risk of heart disease and stroke (Mackay and Mensah 2004). It is a well-known fact that high serum total and LDL cholesterol are important risk factors for CAD (Gould et al. 1998; Ballantyne 1998). Elevated LDL levels are associated with myocardial infarction and peripheral vascular disease, while HDL offers protection against these disorders.

The descriptive statistics regarding the total cholesterol and triglycerides in various cardiovascular disorders has been summarized in Tables 7.1 and 7.20, respectively. A graphical comparison of total cholesterol, HDL cholesterol, LDL cholesterol, and triglyceride in angina patients with control has been given in Figs. 9.1, 9.2, 9.3, and 9.4, respectively. It was interesting to note that the mean level of total cholesterol in all cardiovascular disorders has been in the normal limit of up to 200 mg/dL (Table 6.25). However, about 68–78 % of the studied cases had normal total cholesterol and 22–32 % had total cholesterol higher than 200 mg/dL (Table 7.28). Similarly, in the case of TG, 13–36 % of the study samples had elevated triglycerides, i.e., above 200 mg/dL (Fig. 9.5).

These findings are partially in line with those of Njelekela et al. (2001) who reported from Tanzania 25 % prevalence of elevated serum total cholesterol (cholesterol >5.2 mmol/L) and 15 % prevalence of elevated triglycerides (TG $\geq$ 1.7 mmol/L), among obese adults with CVD disorders over 35 years of age, with women affected more than men (Swai et al. 1993). The descriptive statistics of prevalence of total cholesterol, HDL cholesterol, and LDL cholesterol in angina patients have been summarized in Tables 7.1, 7.9, and 7.13, respectively. Total cholesterol, LDL cholesterol, HDL cholesterol, and TG were normal in 77, 70, 52, and 86 % of the study sample, respectively (Table 7.28).

N.B. Rizvi and S.A. Nagra, *Minerals and Lipids Profiles in Cardiovascular Disorders in South Asia*, DOI 10.1007/978-3-642-34249-3_9,

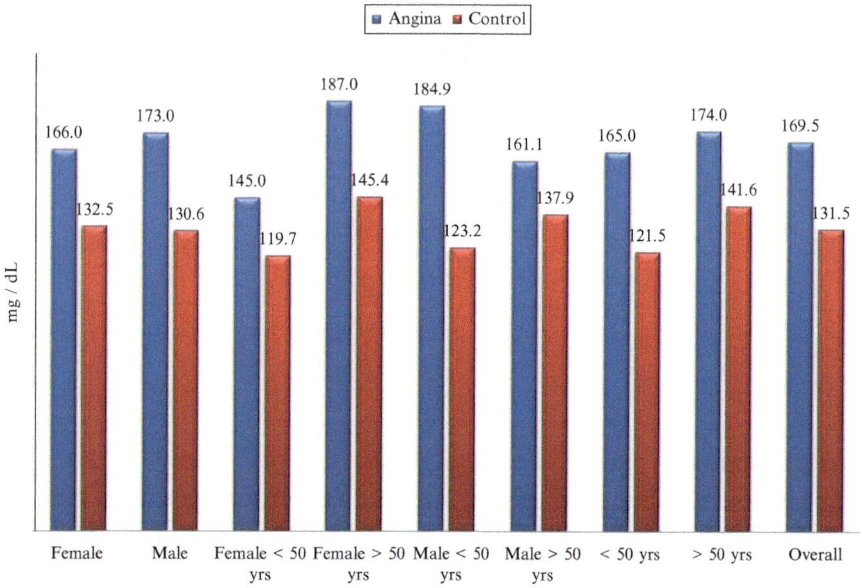

Fig. 9.1 Comparison of mean level of total cholesterol (mg/dL) in angina patients and control group

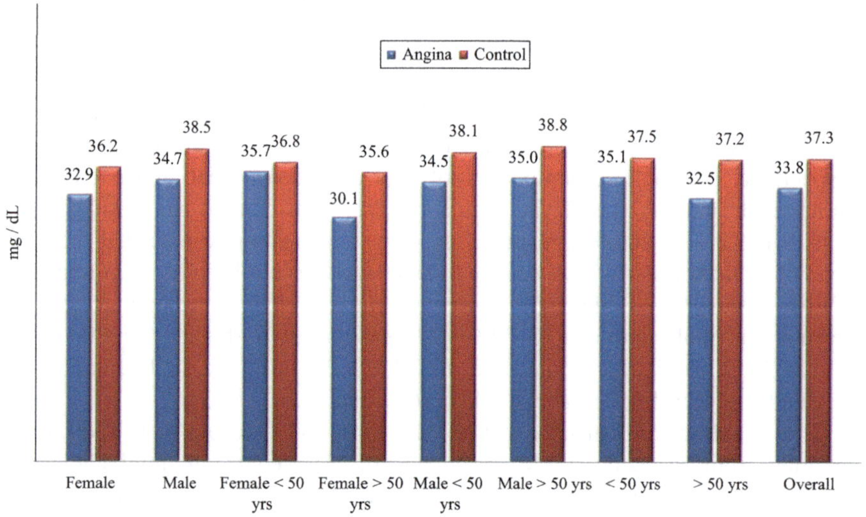

Fig. 9.2 Comparison of mean level of HDL cholesterol (mg/dL) in angina patients and control group

Total cholesterol, LDL-cholesterol, and triglyceride (TG) levels were found to be associated with age and genders in individuals having CVD disorders (Tables 7.2, 7.13, and 7.21, respectively). Total cholesterol, LDL cholesterol, and

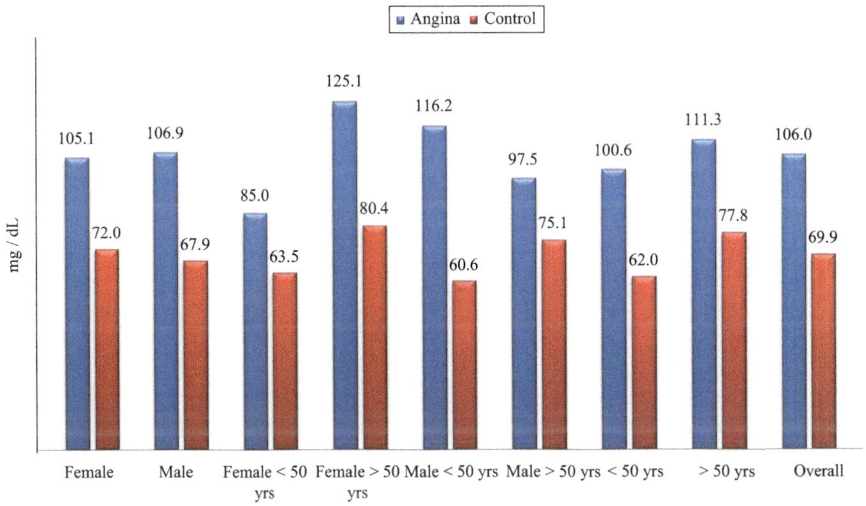

Fig. 9.3 Comparison of mean level of LDL cholesterol (mg/dL) in angina patients and control group

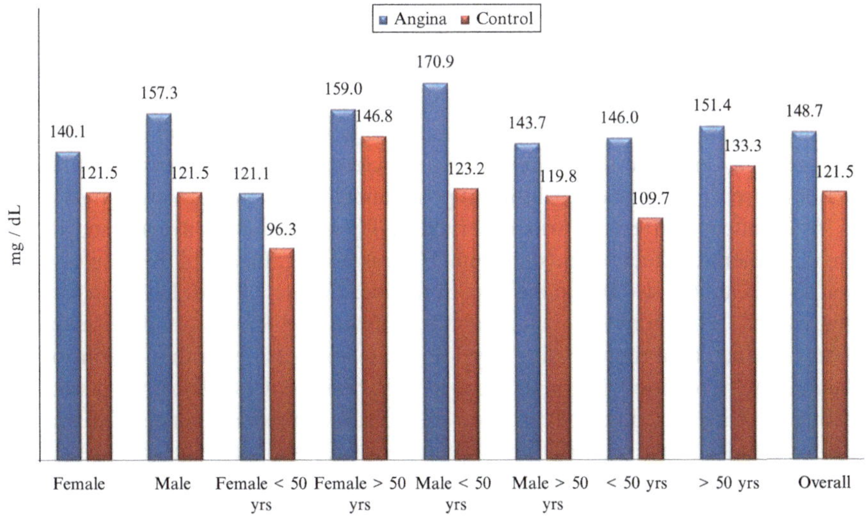

Fig. 9.4 Comparison of mean level of triglyceride (mg/dL) in angina patients and control group

triglycerides were significantly high in angina patients when compared with the control (Tables 7.8, 7.19, and 7.27, respectively). A nonsignificant difference was observed in HDL cholesterol in angina patients when compared with the control (Table 7.12). The findings of our study showed that in angina patients, there were lipid abnormalities, with the exception of HDL cholesterol. It was also observed that total cholesterol and LDL cholesterol increased with age in female angina

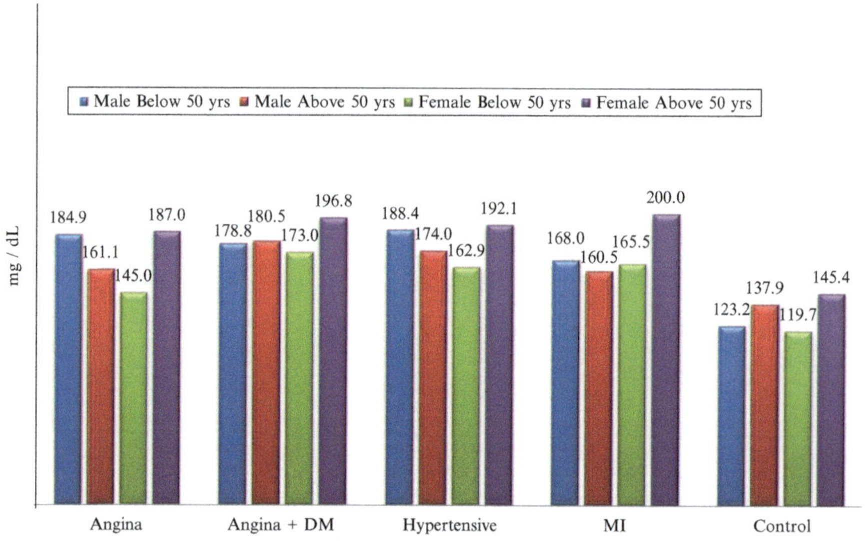

Fig. 9.5 Gender- and age-wise comparison of mean level of total cholesterol (mg/dL) in different study group

patients (187 ± 31.76, 125 ± 35.34), whereas, HDL cholesterol decreased with age in females. Reduction in HDL cholesterol after menopause was considered to be associated with estrogen deficiency (Miller and Miller 1975). Menopause appeared to be associated with adverse changes in blood lipid profile. These changes may enhance the process of atherosclerosis, which is considered to be a major cause of death and disability in postmenopausal women (Newnham 1993).

A significant difference was observed in TG levels between male and female angina patients. Effect of age on serum TG was nonsignificant in the case of males and significant ($P < 0.05$) in the case of females (Table 7.21). It has been reported that risk of CHD increases with age both in males and females, and it is expected that some factors of reproductive physiology are responsible for this. Hokanson and Austin (1996) reported that an increase in triglycerides of 1 mmol/L was associated with a 76 % increased risk of cardiovascular disease in women versus 32 % in men. Each 1 mg/dL decrease in HDL cholesterol has been shown to increase the risk for CAD by 2 % and 3 % in men and women, respectively (Gordon et al. 1989). Our results do not agree with those of Gururajan et al. (2010) who reported that TC levels were nonsignificantly elevated in unstable angina when compared with controls. Triglyceride levels increased though nonsignificantly in the patient groups than controls. These differences may be attributed to the geographical deviation of the two study samples and variation in other environmental factors, mainly nutrition. From the foregoing discussion it can be inferred that mere information about lipid profile of local population should not constitute the basis of fitness or pathogenesis about angina. This information can serve merely as a support for the adoption of any prophylactic measure.

9.2 Angina with Coexistent Diabetes Mellitus

Diabetes mellitus arises when insufficient insulin is produced or when the available insulin does not function correctly. Diabetes accounts for 6 % of the total global mortality, with 50 % diabetes-associated deaths being attributed to cardiovascular diseases (International Diabetes Federation 2008). Hyperlipidemia, as a metabolic abnormality, is frequently associated with diabetes mellitus. It is well known that patients with diabetes have two- to threefold increased risk of developing CAD (Haffner et al. 1998). It has been estimated that there are more than 154 million diabetics worldwide, and its prevalence is on the increase in the developing countries (Bennett et al. 2000). Atherogenic dyslipidemia in diabetic patients is often called as diabetic dyslipidemia. Diabetic dyslipidemia is characterized by high triglycerides, low HDL, and normal LDL cholesterol in most of the patients. A significant effect ($P < 0.05$) of age and gender was observed on total cholesterol (Table 7.2), LDL-cholesterol (Table 7.14), HDL-cholesterol (Table 7.10), and triglyceride (Table 7.21) levels in angina patients with coexistent DM.

Comparison of TC (Table 7.8), LDL cholesterol (Table 7.19), HDL cholesterol (Table 7.12), and TG (Table 7.27) with control indicated a significant ($P < 0.05$) increase in TC, HDL-cholesterol, and TG levels and a significant ($P < 0.05$) decrease in LDL-cholesterol level. A graphical comparison of total cholesterol, LDL cholesterol, HDL cholesterol, and triglyceride in angina with coexistent DM patients with control has been given in Figs. 9.6, 9.7, 9.8, and 9.9, respectively. An increase in serum total lipids, total cholesterol, and serum triglycerides has been reported in diabetic subjects as compared to normal controls. An increase in total cholesterol and LDL-cholesterol levels was observed in postmenopausal females, i.e., above 50, having angina with coexistent DM. These results are in line with those of Kanaley et al. (2001) who reported impact of age and menopausal status on fat distribution. However, our results are partially in variant to those of Mengesha (2006) who reported an increase in LDL-cholesterol levels in female diabetic patients and found no significant difference in other plasma lipid levels between males and females. An elevation in serum triglyceride levels was observed in males above 50 years of age. Females less than 50 showed a slight increase in TG levels. These results partially agree with a cross-sectional survey from South Africa which reported that in type II diabetes, 24–25 % men and 17–18 % women showed an increase in serum triglyceride levels (Nthangeni et al. 2002). There were no relevant age-related variations in HDL-cholesterol levels in both genders except males below 50, which showed very low levels of HDL cholesterol (Table 7.18). In diabetic subjects, gender plays a significant effect on risk of CHD. Significant difference in lipid profile of male and female diabetics may be attributed to sex hormones that play a unique role for lipid metabolism (Smith and Lall 2008). 31.7 % angina patients with coexistent DM had cholesterol levels above 200 mg/dL (Table 65). In angina patients with coexistent DM, it was observed that 26.7 % had LDL-cholesterol levels below 130 mg/dL, 35.8 % had TG greater or equal to 200 mg/dL, and 45.8 % were found to have HDL-cholesterol levels between 35 and 65 mg/dL (Table 7.28).

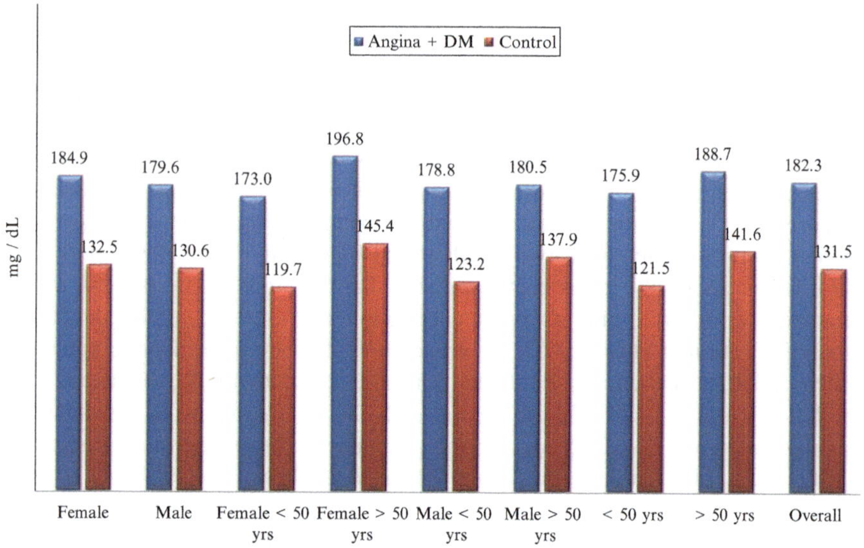

Fig. 9.6 Comparison of mean level of total cholesterol (mg/dL) in angina patients with coexistent DM and control group

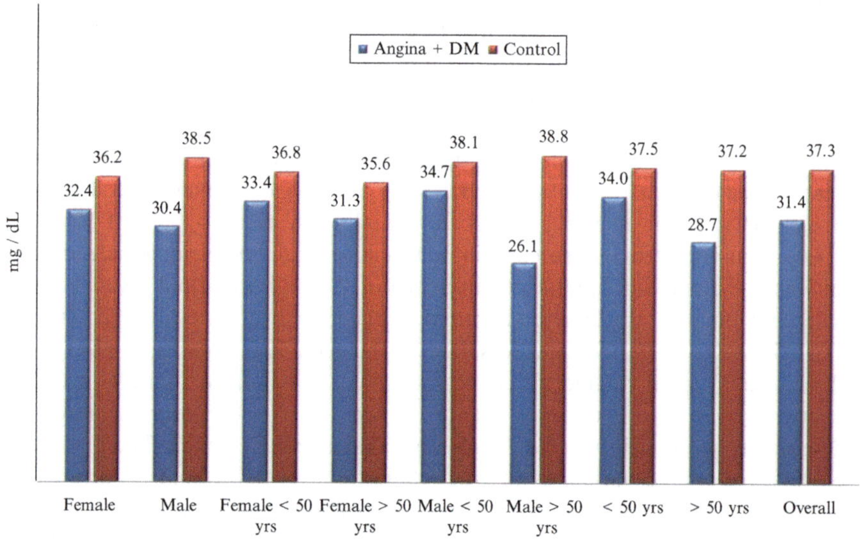

Fig. 9.7 Comparison of mean level of HDL cholesterol (mg/dL) in angina patients with coexistent DM and control group

Among patients with diabetes, the American Diabetes Association recommends that HDL cholesterol be above 40 and 50 mg/dL for men and women, respectively (Haffner 2004). It is generally believed that HDL cholesterol bears an inverse

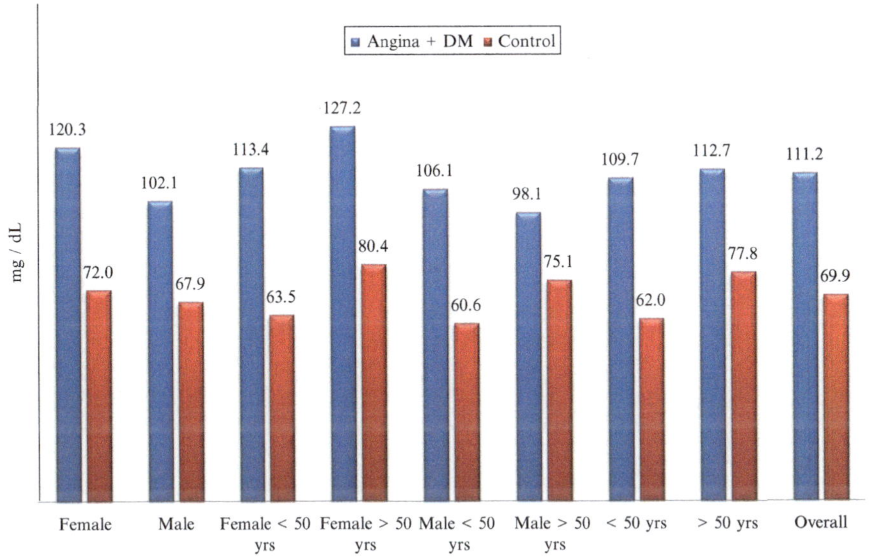

Fig. 9.8 Comparison of mean level of LDL cholesterol (mg/dL) in angina patients with coexistent DM and control group

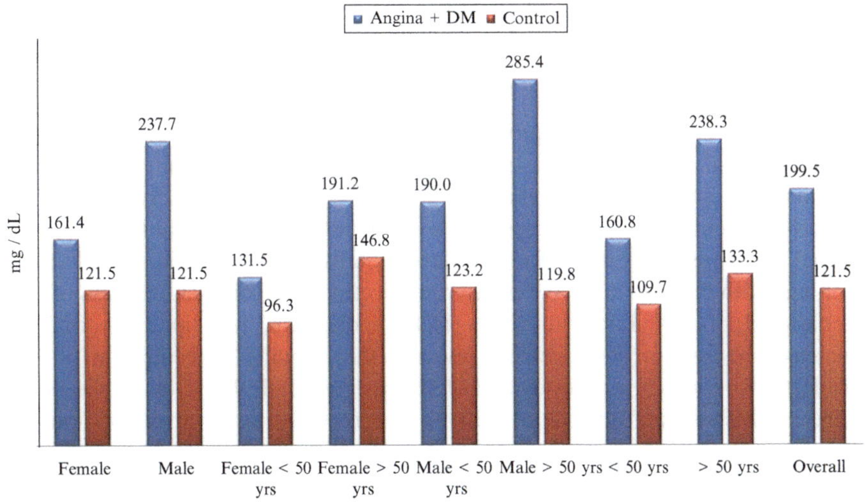

Fig. 9.9 Comparison of mean level of triglyceride (mg/dL) in angina patients with coexistent DM and control group

relationship to the risk of atherosclerosis and CHD; the higher the level, the smaller the risk (Khoo et al. 1997; Tao et al. 1992). It has also been observed that Asian Indians have low HDL-cholesterol levels, which could be one of the risk factors for premature CAD in this ethnic group (Enas et al. 1996; Beckles et al. 1986).

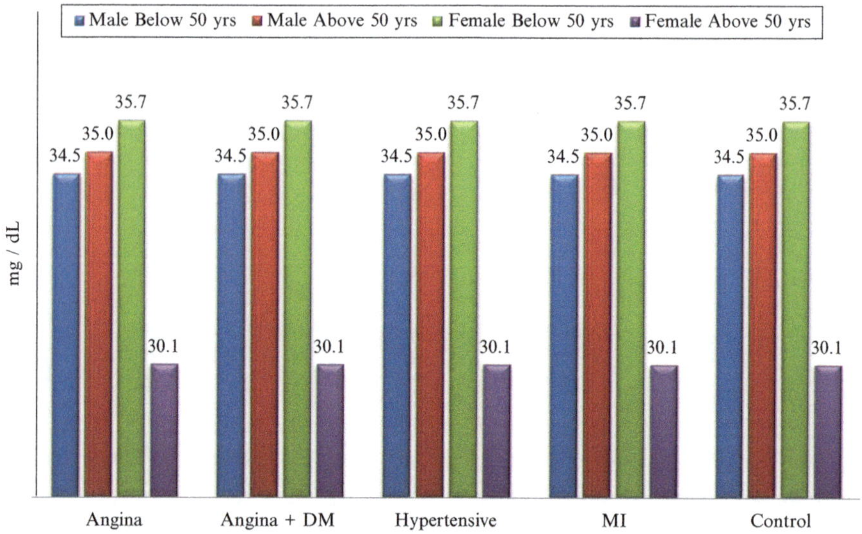

Fig. 9.10 Gender- and age-wise comparison of mean level of HDL cholesterol (mg/dL) in different study group

Findings of the present study agree with those of Habib (2006) who reported a high prevalence of undesirable level of LDL cholesterol in Pakistani diabetic subjects. In a study of diabetic patients in Ethiopia, 18.5 % and 14.2 % were found to have hypercholesterolemia and hypertriglyceridemia, respectively, and hypercholesterolemia was found in a high proportion in females and type II diabetic patients (Seyoum et al. 2003). According to the third US National Health and Nutrition Examination Survey and the Behavioral Risk Factors Surveillance System, 58 % of diabetic patients had LDL-cholesterol levels greater than or equal to 130 mg/dL, which was considered a risk for CHD and atherosclerosis (Saaddine et al. 2002). Patients with diabetes can have many complications including an increase of LDL cholesterol, TG, and low levels of HDL cholesterol, and lipid abnormalities increase the risk of atherogenicity even if the absolute concentration of LDL cholesterol is insignificantly increased (Haffner et al. 1998). Our results do not agree with those of Ugwu et al. (2009) who reported lower levels of lipid and lipoprotein profiles in diabetics when compared with controls. They also observed no gender difference in the lipid metabolism between the diabetic and nondiabetic males and females. Since, lipid profile is affected by a number of factors such as age, gender, racial differences, dietary habits, and socioeconomic status, geographical conditions, even in the case of normal and healthy individuals, the apparent differences may be attributed to these factors. In spite of the ethnic and cultural differences, it was observed that diabetics have significantly higher prevalence of dyslipidemia (Bermudez et al. 2002) (Fig. 9.10).

9.3 Hypertension

Hypertension has been recognized as most common cardiovascular disorder (Reports of the Joint National Committee 1997) and a leading cause of morbidity and mortality in both developed and developing countries. Hypertension in adults is arbitrarily defined as systolic pressure to or greater than 160 mmHg and/or diastolic pressure equal to or greater than 95 mmHg. It has been observed that blood lipids and lipoproteins are closely associated with hypertension. The association between hypertension and dyslipidemia is well established, and both may add up to increase patients' susceptibility to the development of CHD. Dyslipidemia that causes endothelial dysfunction may lead to hypertension (Oparil et al. 2003). Analysis of variances indicated a significant ($P < 0.05$) effect of age and gender on total cholesterol (Table 7.2), LDL cholesterol (Table 7.14), and triglycerides (Table 7.21). A graphical comparison of total cholesterol, LDL cholesterol, HDL cholesterol, and triglyceride in hypertensive patients with control has been given in Figs. 9.11, 9.12, 9.13, and 9.14, respectively. An increase in TC and LDL-cholesterol levels was observed in postmenopausal female hypertensive patients. TG levels were higher in the case of elderly (above 50 years) males; however, HDL cholesterol did not show a significant variation between hypertensive males and females of all age groups. These findings are in line with the work of Idemudia and Ugwuja (2009) who demonstrated that the prevalence of hypertension is highest in age groups between 40–49 years and 50–59 years for males and females, respectively. It can also be observed that the abnormalities in lipid profile increased with increasing age in hypertensive persons, which are in line with the work of Saha et al. (2006) who reported that the mean value of serum TC, TG, and LDL cholesterol was significantly ($P < 0.05$) higher in hypertensive patients, with significantly lower HDL-cholesterol levels when compared with the control group. It was further reported that TC, TG, and LDL-cholesterol levels were positively associated with hypertension, whereas HDL cholesterol had no significant changes with hypertension. Our results are further supported by Youmbissi et al. (2001) and Mgonda et al. (1998), who observed significantly higher ($P < 0.05$) plasma total cholesterol, triglycerides, and LDL-cholesterol levels in the hypertensive than in the normotensive patients. Several studies have consistently showed a positive relationship between age and blood pressure in both developed and developing countries (Singh et al. 1997; Whelton 1994) (Fig. 9.15).

Some researchers have argued that the desired range of plasma total cholesterol concentrations as advocated for developed countries may have to be reviewed for developing countries because subjects in developing countries could be prone to developing CHD at a lower plasma cholesterol level (Singh et al. 1998). 28.3 % hypertensive patients were observed to have TC levels above 200 mg/dL. 12.5 % cases had TG levels higher than ≥200 mg/dL, 37.5 % had LDL-cholesterol levels ≥130 mg/dL., 50.8 % had HDL-cholesterol levels below 35 mg/dL, and 49.2 % had HDL-cholesterol levels between 35 and 65 mg/dL (Table 7.28). Individuals with high blood TC levels have a higher prevalence of hypertension, and those with high

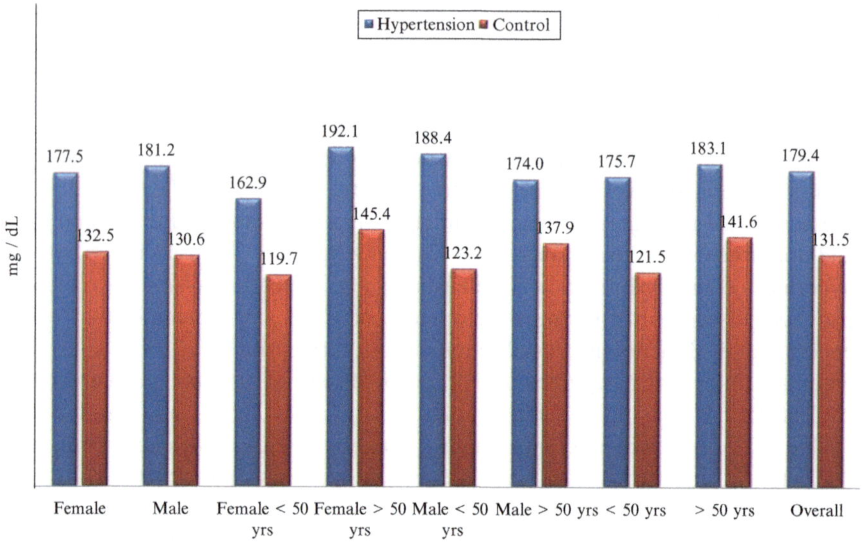

Fig. 9.11 Comparison of mean level of total cholesterol (mg/dL) in hypertensive patients and control group

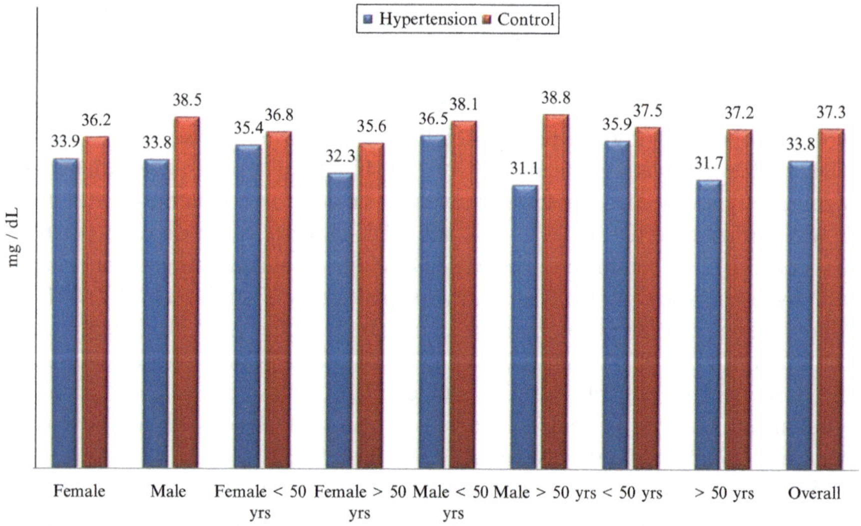

Fig. 9.12 Comparison of mean level of HDL cholesterol (mg/dL) in hypertensive patients and control group

blood pressure have a higher prevalence of hypercholesterolemia (Johnson et al. 2004; European Society of Hypertension 2003). Gaziano et al. (1999) observed a potential interaction between elevated TC and hypertension in the development of myocardial infarction also. According to expert panel of National Cholesterol

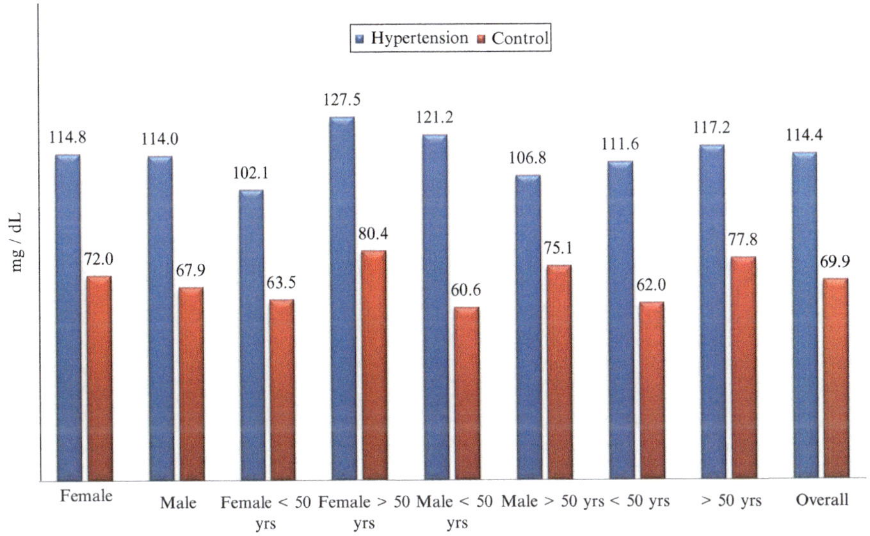

Fig. 9.13 Comparison of mean level of LDL cholesterol (mg/dL) in hypertensive patients and control group

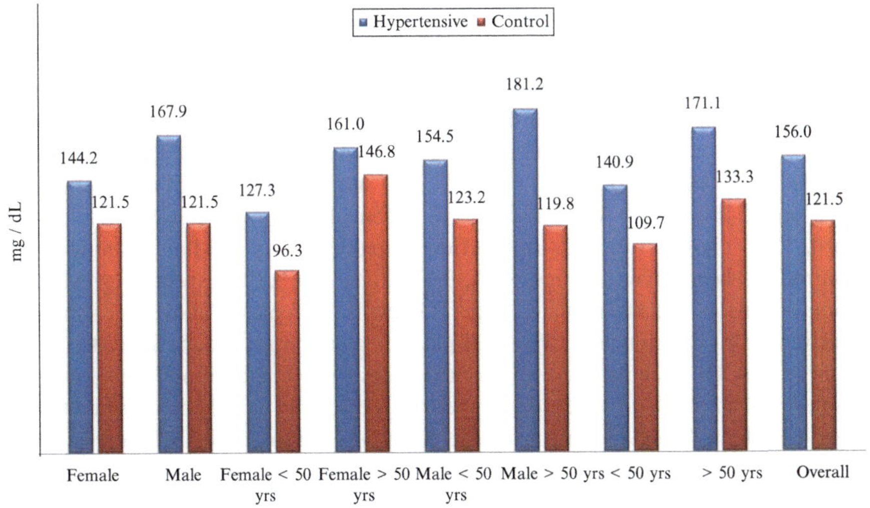

Fig. 9.14 Comparison of mean level of triglyceride (mg/dL) in hypertensive patients and control group

Education Program, the practical action points for considering an association between lipid profile levels with the risk of developing CAD were calculated, with TC > 200 mg/dL, TG > 150 mg/dL, LDL cholesterol > 130 mg/dL, and HDL cholesterol < 40 mg/dL as risk factors in CAD (Haffner et al. 1996).

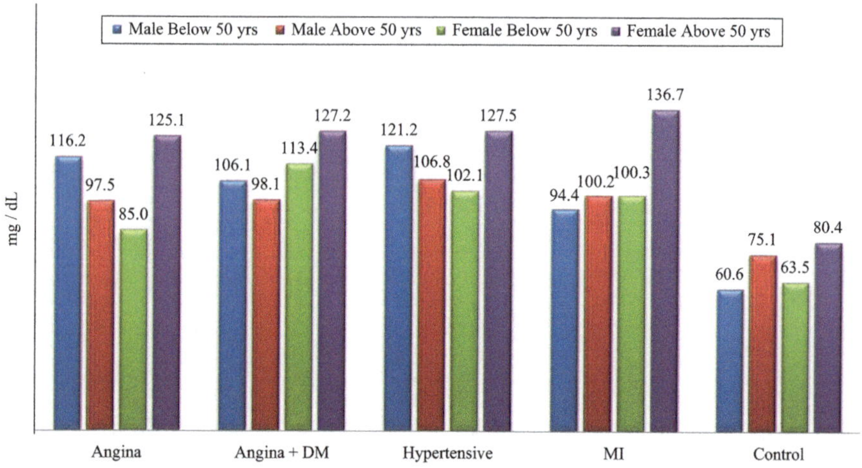

Fig. 9.15 Gender- and age-wise comparison of mean level of LDL cholesterol (mg/dL) in different study group

9.4 Myocardial Infarction

Myocardial infarction is a terminal consequence of CHD. The basic cause for CHD is atherosclerosis, when arteries become narrow or hardened due to cholesterol plaque buildup. Myocardial infarction occurs when a coronary artery is so severely blocked that there is a significant reduction or break in the blood supply, causing damage or death to a portion of the myocardium. Analyses of variance indicated a significant ($P < 0.05$) effect of age and gender on TC (Table 7.2), LDL cholesterol (Table 7.14), HDL cholesterol (Table 7.10), and TG (Table 7.21), in patients of myocardial infarction. A graphical comparison of total cholesterol, HDL cholesterol, LDL cholesterol, and triglyceride in patients' myocardial infarction with control has been given in Figs. 9.16, 9.17, 9.18, and 9.19, respectively. Multiple comparisons of means revealed that TG levels in females above 50 showed a nonsignificant difference between MI patients and control group. Same trend was followed by HDL-cholesterol levels in females (above and below 50). Low HDL cholesterol and high prevalence of hypertriglyceridemia have been reported to play a major role in the development of atherosclerosis in Pakistani AMI patients (Iqbal et al. 2004). It is evident from Tables 7.13 and 7.21 that postmenopausal females had the highest levels of TC and LDL cholesterol, i.e., 200 ± 22.96 mg/dL and 136.70 ± 29.99 mg/dL as compared to premenopausal females and males of same age group. Premenopausal women are said to be protected against CHD, but this protection is lost once the women become postmenopausal (Hong et al. 1992). Menopause appears to be associated with adverse changes in blood lipid profile. Serum triglyceride levels were higher in younger males (i.e., below 50 years), whereas there was no variation in TG levels in females of both age groups.

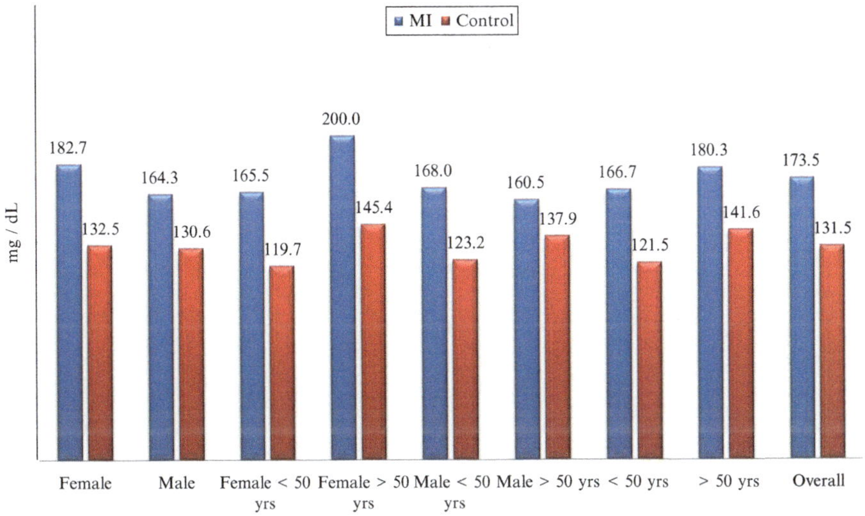

Fig. 9.16 Comparison of mean level of total cholesterol (mg/dL) in MI patients and control group

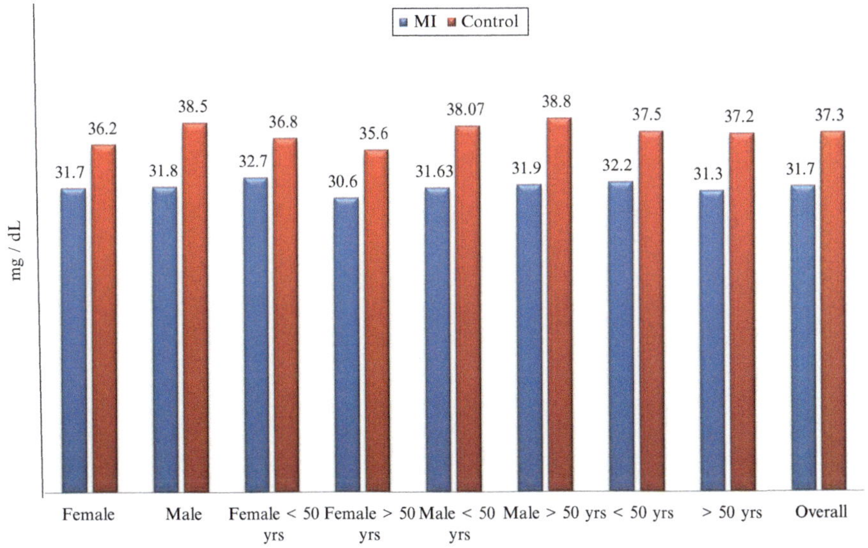

Fig. 9.17 Comparison of mean level of HDL cholesterol (mg/dL) in MI patients and control group

Triglyceride levels were least affected by age in the case of females. Wenger (2003) reported that high levels of TG are associated with a greater risk of CHD in women than men. Kumar and Sivakanesan (2009) observed significantly higher (129 mg/dL) triglyceride (TG) values in MI patients when compared with controls (107.8 mg/dL). There was no age- and gender-related variation in HDL-cholesterol

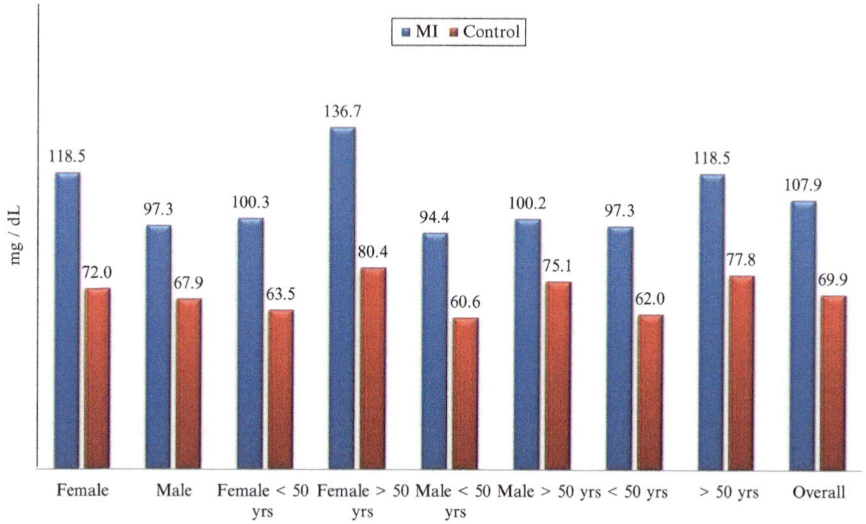

Fig. 9.18 Comparison of mean level of LDL cholesterol (mg/dL) in MI patients and control group

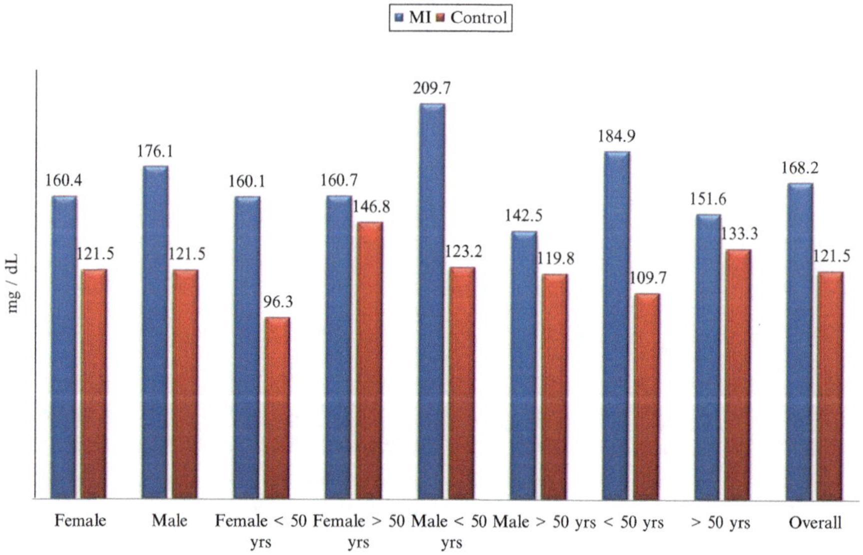

Fig. 9.19 Comparison of mean level of triglycerides (mg/dL) in MI patients and control group

levels in MI patients and found to be lower as compared with the control group. Kumar and Sivakanesan (2009) also observed significantly lower HDL-cholesterol values (41.3 mg/dL) in MI patients as compared to the control (50.5 mg/dL). It was observed that low concentrations of HDL cholesterol were a better predictor of

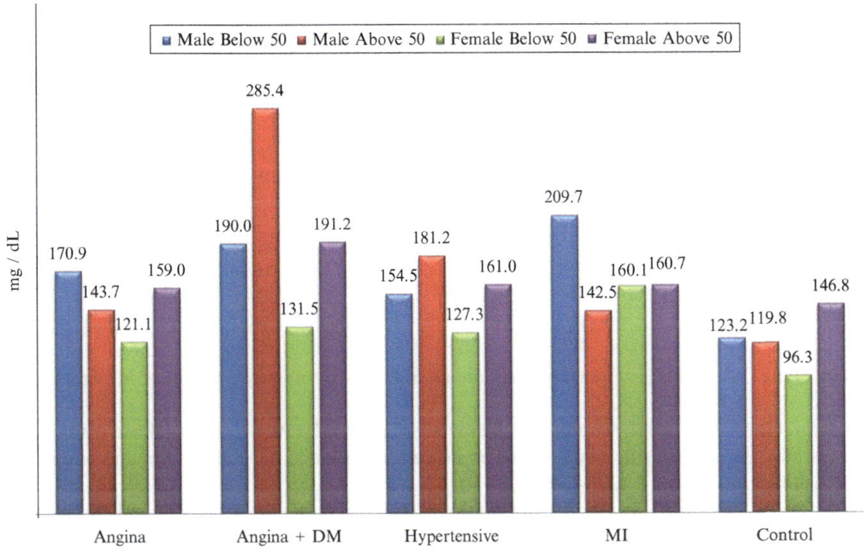

Fig. 9.20 Gender- and age-wise comparison of mean level of triglyceride (mg/dL) in different study group

coronary risk than high concentrations of LDL cholesterol, especially in the case of females (Wenger 2003) (Fig. 9.20).

Twenty-five percent MI patients had TC levels $\geq$ 200 mg/dL. TG levels $\geq$ 200 mg/dL were observed in 22.5 % MI patients. Thirty percent MI patients had LDL-cholesterol levels $\geq$ 130 mg/dL and 45 % had HDL-cholesterol levels between 35 and 65 mg/dL (Table 65). According to the National Cholesterol Education Program, a low HDL cholesterol is defined as a level less than 40 mg/dL (NCEP 2001). More than 50 % MI patients had HDL-cholesterol levels below 35 mg/dL. American Heart Association has recently concluded that in women, HDL is low when it is below 50 mg/dL (Mosca et al. 2004).

In general, the higher the HDL cholesterol, the greater is the capacity to remove cholesterol and prevent dangerous blockages from developing in the blood vessels. Karthikeyan et al. (2009) observed a nonsignificant tendency toward weaker protection afforded by normal or high levels of HDL cholesterol among south Asians suffering from AMI. On the contrary, it has been reported that no significant change was observed in serum total lipid and TC in MI patients (Nigam et al. 2004). Tsimikas et al. (2003) also observed no significant difference in lipid profile levels between control and AMI group. From these studies it is clear that phase changes occur in patients following MI, and therefore, it is recommended that for the detection of hyperlipidemia in MI patients, serum lipids should be assessed either within 24 h after infarction or after 2–3 months (Nigam et al. 2004). Vetter et al. (1974) recorded a progressive fall in triglyceride levels in the second hour after MI, but Ryder et al. (1984) showed no significant difference.

Lipid profile is generally considered as an indicator for the tendency to develop cardiovascular problems. It is interesting to note that in Pakistani population whether control or cardiovascular patients, the lipid profile falls in the normal range. Though, total cholesterol and triglycerides were generally high in cardiovascular patients than control. However, both fell in the normal range. It may be postulated that lipid profile should not be considered a real indicator to predict the possibility of the onset of CVD disorder.

References

Ballantyne CM (1998) Low-density lipoproteins and risk for coronary artery disease. Am J Cardiol 82:3Q–12Q

Beckles GL, Miller GJ, Kirkwood BR, Alexis SD, Carson DC (1986) High total and cardiovascular disease mortality in adults of Indian descent in Trinidad, unexplained by major coronary risk factors. Lancet 1:1298–1301

Bennett PH, Lekoith D, Taylor SI, Olefsky JM (2000) Epidemiology of type II diabetes mellitus. In: Lekoith D, Taylor SI, Olefsky JM (eds) Diabetes mellitus, 2nd edn. Wolter, New York, pp 544–557

Bermudez OI, Velez-Carrasco W, Schaefer EJ, Tucker KL (2002) Dietary and plasma lipid, lipoprotein, and apolipoprotein profiles among elderly Hispanics and non-Hispanics and their association with diabetes. Am J Clin Nutr 76(6):1214–1221

Enas EA, Garg A, Davidson MA (1996) Coronary heart disease and its risk factors in first-generation immigrant Asian Indians to the United States of America. Indian Heart J 48:343–352

European Society of Hypertension-European Society of Cardiology Guidelines Committee (2003) 2003 European Society of Hypertension-European Society of Cardiology guidelines for the management of arterial hypertension. J Hypertens 21:1011–1053

Expert Panel on Detection, Evaluation, and Treatment of High Blood Cholesterol in Adults (2001) Executive Summary of the Third Report of the National Cholesterol Education Program (NCEP) Expert Panel on Detection, Evaluation, and Treatment of High Blood Cholesterol in Adults (Adult Treatment Panel III). JAMA 285:2486–2497

Gaziano JM, Sesso HD, Breslow JL (1999) Relation between systemic hypertension and blood lipids on the risk of myocardial infarction. Am J Cardiol 84:768–773

Gordon DJ, Probstfield JL, Garrison RJ, Neaton JD, Castelli WP, Knoke JD (1989) High-density lipoprotein cholesterol and cardiovascular disease. Four prospective American studies. Circulation 79:8–15

Gould AL, Rossouw JE, Santanello NC, Heyse JF, Furberg CD (1998) Cholesterol reduction yields clinical benefit: impact of statin trials. Circulation 97:946–952

Gururajan P, Gurumurthy P, Nayar P, Chockalingam M, Bhuvaneshwari S, Babu S, Sarasabharati A, Victor D, Cherian KM (2010) Lipid profile and non-enzymic antioxidant status in patients with acute coronary syndrome in South India. Heart Lung Circ 19(2):75–80

Habib SS (2006) Frequency distribution of atherogenic dyslipidemia in Saudi type 2 diabetic patients. Pak J Physiol 2:20–23

Haffner SM, American Diabetes Association (2004) Dyslipidemia management in adults with diabetes. Diabetes Care 27:S68–S71

Haffner SM, Miettinen H, Gaskill SP, Stern MP (1996) Metabolic precursors of hypertension: the San Antonio Heart Study. Arch Intern Med 156:1994–2000

Haffner SM, Lehto S, Ronnemaa T, Pyorala K, Laakso M (1998) Mortality from coronary heart disease in subjects with type 2 diabetes and in nondiabetic subjects with and without prior myocardial infarction. N Engl J Med 339:229–234

Hokanson JE, Austin MA (1996) Plasma triglyceride level is a risk factor for cardiovascular disease independent of high-density lipoprotein cholesterol level: a meta-analysis of population-based prospective studies. J Cardiovasc Risk 3:213–219

Hong MK, Romm PA, Reagan K, Green CE, Rackley CE (1992) Effects of oestrogen replacement therapy on serum lipid values and angiographically defined coronary artery disease in post-menopausal women. Am J Cardiol 69:176–178

Idemudia JO, Ugwuja EI (2009) Plasma lipid profiles in hypertensive Nigerians. Internet J Cardiovasc Res 6(2):8

International Diabetes Federation (2008) Diabetes prevalence [article online]. Available from http://www.idf.org/home/index.cfm?node=264

Iqbal MP, Shafiq M, Mehboobali N, Iqbal SP, Abbasi K (2004) Variability in lipid profile in patients with acute myocardial infarction from two tertiary care hospitals in Pakistan. J Pak Med Assoc 54(11):544–549

Johnson ML, Pietz K, Battleman DS, Beyth RJ (2004) Prevalence of comorbid hypertension and dyslipidemia and associated cardiovascular disease. Am J Manag Care 10:926–932

Joint National Committee on Detection, Evaluation, and Treatment of High Blood Pressure (JNC VI) (1997) The sixth reports of the Joint National Committee on Detection, Evaluation, and Treatment of High Blood Pressure (JNC VI). Arch Intern Med 157:2413–2446

Kanaley JA, Sames C, Swisher L, Swick AG, Steppan CM, Feiglin D, Jaynes EB, Meyer RA, Weinstock RS (2001) Abdominal fat distribution in pre and post menopausal women. The impact of physical activity, age and menopausal status. Metabolism 50(8):976–982

Karthikeyan G, Teo KK, Islam S, McQueen MJ, Pais P, Wang X, Sato H, Lang CC, Sitthi-Amorn C, Pandey MR, Kazmi K, Sanderson JE, Yusuf S (2009) Lipid profile, plasma apolipoproteins, and risk of a first myocardial infarction among Asians: an analysis from the INTERHEART Study. J Am Coll Cardiol 53(3):244–253

Khoo KL, Tan H, Leiw YM (1997) Serum lipids and their relationship with other coronary risk factors in healthy subjects in a city clinic. Med J Malaysia 52:38–52

Kumar A, Sivakanesan R (2009) Serum lipid profile abnormality in predicting the risk of myocardial infarction in elderly normolipidaemic patients in south Asia: a case-controlled study. Internet J Altern Med 6:2

Mackay J, Mensah G (2004) Atlas of heart disease and stroke. World Health Organization, Geneva

Mengesha AY (2006) Lipid profile among diabetes patients in Gaborone, Botswana. S Afr Med J 96:147–148

Mgonda YM, Ramaiya KL, Swai ABM, Mc-Larty DG, George KM, Alberti M (1998) Insulin resistance and hypertension in non-obese Africans in Tanzania. Hypertension 31:114–118

Miller GJ, Miller NE (1975) Plasma high density lipoprotein concentration and development of ischemic heart disease. Lancet 1:16–19

Mosca L, Appel LJ, Benjamin EJ, Berra K, Chandra-Strobos N, Fabunmi RP, Grady D, Haan CK, Hayes SN, Judelson DR, Keenan NL, McBride P, Oparil S, Ouyang P, Oz MC, Mendelsohn ME, Pasternak RC, Pinn VW, Robertson RM, Schenck-Gustafsson K, Sila CA, Smith SC Jr, Sopko G, Taylor AL, Walsh BW, Wenger NK, Williams CL, American Heart Association (2004) Evidence-based guidelines for cardiovascular disease prevention in women. Circulation 109:672–693

Newnham HH (1993) Oestrogens and atherosclerotic vascular disease-lipid factors. Bailleres Clin Endocrinol Metab 7:61–93

Nigam PK, Narain VS, Hasan M (2004) Serum lipid profile in patients with acute myocardial infarction. Indian J Clin Biochem 19(1):67–70

Njelekela M, Negishi H, Nara Y, Tomohiro M, Kuga S, Noguchi T, Kanda T, Yamori M, Mashalla Y, Jian Liu L (2001) Cardiovascular risk factors in Tanzania: a revisit. Acta Trop 79 (3):231–239

Nthangeni G, Steyn NP, Alberts M (2002) Dietary intake and barriers to dietary compliance in black type II diabetic patients attending primary health-care services. Public Health Nutr 5:329–338

Oparil S, Zaman MA, Calhoun DA (2003) Pathogenesis of hypertension. Ann Intern Med 139:761–776

Ryder REJ, Hayes TM, Mulligan IP, Kingwood JC, Williams S, Owens DR (1984) How soon after myocardial infarction should plasma lipid values be assessed. Br Med J 289:1651–1653

Saaddine JB, Engelgau MM, Beckles GL (2002) A diabetes report card for the United States: quality of care in the 1990s. Ann Intern Med 136:565

Saha MS, Sana NK, Shaha RK (2006) Serum lipid profile of hypertensive patients in the Northern region of Bangladesh. J Biosci 14:93–98

Seyoum B, Abdulkadir J, Berhanu P (2003) Analysis of serum lipids and lipoproteins in Ethiopian diabetic patients. Ethiop Med J 41(1):1–8

Singh RB, Beegom R, Ghosh S, Niaz MA, Rastogi V (1997) Epidemiological study of hypertension and its determinants in an urban population of North India. J Hum Hypertens 11:679–685

Singh RB, Niaz MA, Rastogi SS, Bajaj S, Gaoli Z, Shoumin Z (1998) Current zinc intake and risk of diabetes and coronary artery disease and factors associated with insulin resistance in rural and urban populations of North India. J Am Coll Nutr 17(6):564–570

Smith S, Lall AM (2008) A study on lipid profile levels of diabetics and non-diabetics among Naini region of Allahabad, India. Turk J Biochem 33(4):138–141

Swai AB, McLarty DG, Kitange HM, Kilima PM, Tatalla S, Keen N, Chuwa LM, Alberti KG (1993) Low prevalence of risk factors for coronary heart disease in rural Tanzania. Int J Epidemiol 22(4):651–659

Tao S, Li Y, Xiao Z, Cen R, Zhang H, Zhuo B, Chen P, Liao Y (1992) Serum lipids and their correlates in Chinese urban and rural population of Beijing and Guangzhou. PRC-USA Cardiovascular and Cardiopulmonary Epidemiology Research Group. J Epidemiol 21:893–903

Tsimikas S, Bergmark C, Beyer RW (2003) Temporal increases in plasma markers of oxidized low density lipoprotein strongly reflect the presence of acute coronary syndrome. J Am Coll Cardiol 41:360–370

Ugwu CE, Ezeanyika LUS, Daikwo MA, Amana R (2009) Lipid profile of a population of diabetic patients attending Nigerian National Petroleum Corporation Clinic, Abuja. Afr J Biochem Res 3(3):066–069

Vetter NJ, Strange RC, Adams W, Oliver MF (1974) Initial metabolic and hormonal response to acute myocardial infarction. Lancet 1(7852):284–289

Wenger NK (2003) Coronary heart disease: the female heart is vulnerable. Prog Cardiovasc Dis 46:199–229

Whelton PK (1994) Epidemiology of hypertension. Lancet 334:101–106

Youmbissi TJ, Djoumessi S, Nouedoui C (2001) Profile lipidique d'un group d'hypertendus camerounais noir Africains. Med Afr Noire 31:114–118

Chapter 10
Summary: Minerals

10.1 Copper

The analysis of variance indicated a significant ($P < 0.05$) effect of age and gender on serum copper of all the study groups except control group. Multiple comparisons of mean serum concentration of copper in hypertensive patients belonging to different age and gender indicated that age did not affect the serum copper level in hypertensive males. Below 50 years gender did not appear to affect the copper concentration in serum. Similarly the serum copper level did not differ significantly in males above 50 years and females less than 50 years. All other comparisons were statistically significant ($P < 0.05$).

Multiple comparisons of mean serum concentration of copper in patients with myocardial infarction belonging to different age and gender showed that below 50 years of age, gender did not appear to affect the copper concentration in serum. Serum copper values were significantly ($P < 0.05$) high in almost all age groups, gender, and all CV disorders when compared with the control.

10.2 Magnesium

The statistical appraisal of the data indicated a significant ($P < 0.05$) effect of age and gender within all the study groups and control. Age did not affect the magnesium level in either gender. Similarly, effect of gender was nonsignificant in group below 50 years of age. Multiple comparisons of mean serum concentration of magnesium in angina patients with coexistent DM and hypertensive patients belonging to different age and gender also followed the similar trend. However, multiple comparisons of mean serum magnesium level in MI did not show an appreciable effect of age and gender. Only in below 50 years that the effect of gender was statistically significant ($P < 0.05$)

N.B. Rizvi and S.A. Nagra, *Minerals and Lipids Profiles in Cardiovascular Disorders in South Asia*, DOI 10.1007/978-3-642-34249-3_10,

On an average, in control group, females less than 50 years of age had significantly ($P < 0.05$) higher serum magnesium than males of either age group. Serum magnesium values were significantly ($P < 0.05$) lower in almost all age groups, gender, and all CV disorders when compared with the control.

10.3 Selenium

The analysis of variance indicated a significant ($P < 0.05$) effect of age and gender on serum selenium of angina patients with coexistent DM and control group. A nonsignificant effect of age and gender was observed on serum selenium in the case of patients with angina alone, hypertension, and myocardial infarction. The effect of age on serum selenium was observed nonsignificant in the case of males. Similarly effect of gender on serum selenium was also nonsignificant in below 50 years of age. Strikingly, no appreciable difference was observed in control and different study groups. Females and males less than 50 years of age having angina with coexistent DM and hypertension, respectively, had significantly lower serum selenium than control group. All other comparisons were observed nonsignificant.

10.4 Zinc

Statistical analysis of the data indicated a significant ($P < 0.05$) effect of age and gender within angina and hypertensive patients. Males below 50 years of age had significantly ($P < 0.05$) lower serum zinc than females of the same age group. Males above 50 years of age had significantly ($P < 0.05$) lower serum zinc than females of either age. Effect of age was nonsignificant in the case of hypertensive males. However, females above 50 years of age had significantly ($P < 0.05$) high serum zinc concentration than below 50 years of age. Serum zinc values were significantly ($P < 0.05$) lower in almost all age groups, gender, and all CV disorders when compared with the control.

Chapter 11
Summary: Lipid Profile

11.1 Total Cholesterol

Statistical analysis of the data indicated a significant effect of gender and age on all the study groups. A significant effect ($P < 0.05$) of age on total cholesterol level of both the genders was noted. Females less than 50 years of age had significantly ($P < 0.05$) lower serum total cholesterol level than their male counterparts. Likewise, females above 50 years of age had significantly higher ($P < 0.05$) serum total cholesterol than their male counterparts. Mean cholesterol level in angina patients with coexistent diabetes mellitus indicated a nonsignificant effect of age on total cholesterol level of males and a significant effect ($P < 0.05$) on females. Females above 50 years of age had significantly ($P < 0.05$) higher serum total cholesterol level than males below 50 years of age. All other comparisons were nonsignificant statistically.

Mean cholesterol level in hypertensive patients indicated a nonsignificant effect of age on total cholesterol level of males and a significant effect ($P < 0.05$) on females. Females below 50 years of age had significantly ($P < 0.05$) lower serum total cholesterol level than males below 50 years of age. Similarly females above 50 had significantly ($P < 0.05$) higher serum total cholesterol level than males above 50 years of age. All other comparisons were nonsignificant statistically.

Multiple comparisons of mean cholesterol level in patients of myocardial infarction indicated a nonsignificant effect of age on total cholesterol level of males and a significant effect ($P < 0.05$) on females. Females above 50 years of age had significantly ($P < 0.05$) higher serum total cholesterol level than males of either age group. All other comparisons were nonsignificant statistically. In the case of control group, multiple comparisons of mean indicated that females above 50 years of age had significantly high ($P < 0.05$) level of serum total cholesterol than males and females below 50 years of age. A comparison of total cholesterol level between control and other study indicated that all the study groups had significantly higher ($P < 0.05$) serum total cholesterol level than the control.

N.B. Rizvi and S.A. Nagra, *Minerals and Lipids Profiles in Cardiovascular Disorders in South Asia*, DOI 10.1007/978-3-642-34249-3_11,

11.2 High-Density Lipoprotein Cholesterol

The statistical appraisal of the data indicated a significant ($P < 0.05$) effect of gender and age on high-density lipoprotein cholesterol (HDL cholesterol) in angina patients with coexistent diabetes mellitus. There was a nonsignificant effect of age and gender on other study groups. Multiple comparisons of mean HDL-cholesterol level in angina patients with coexistent diabetes mellitus showed a significant effect ($P < 0.05$) of age on HDL-cholesterol level of males only. Males below 50 years had significantly lower ($P < 0.05$) HDL cholesterol then females of both the age groups.

A comparison of HDL-cholesterol level between control and other study groups indicated that dominantly it was the patients of myocardial infarction who had significantly lower ($P < 0.05$) serum HDL-cholesterol level than the control.

11.3 Low-Density Lipoprotein Cholesterol

The data was subjected to analysis of variance (Table 7.26) which indicated a significant effect of gender and age on all the study groups except control group. Multiple comparisons of mean low-density lipoprotein-cholesterol (LDL-cholesterol) level in angina patients indicated a significant effect ($P < 0.05$) of age on LDL-cholesterol level of females only. Males below 50 years had significantly higher ($P < 0.05$) LDL cholesterol than females of the same age. Females above 50 years of age had significantly ($P < 0.05$) higher LDL-cholesterol level than all the study groups. Multiple comparisons of LDL-cholesterol level in angina patients with coexistent diabetes mellitus revealed that females above 50 years of age had significantly ($P < 0.05$) high LDL cholesterol than males of either age group. Hypertensive females above 50 years of age had significantly ($P < 0.05$) higher LDL cholesterol than males above 50 and females below 50 years of age. Males below 50 years had significantly ($P < 0.05$) higher LDL cholesterol than females of the same age. Almost similar trend of variation in LDL cholesterol was observed in patients of myocardial infarction. A comparison of LDL-cholesterol level between control and other study groups indicated that all the study groups had significantly higher ($P < 0.05$) serum LDL-cholesterol level than the control.

11.4 Triglycerides

Statistical analysis of the data indicated a significant ($P < 0.05$) effect of gender and age on all the study groups. Multiple comparisons of mean serum triglyceride level in angina patients showed a significant effect ($P < 0.05$) of age on serum triglyceride of females only. Males below 50 years had significantly higher

($P < 0.05$) serum triglyceride than females of the same age. Multiple comparisons of serum triglyceride level in angina patients with coexistent diabetes mellitus belonging to different age and gender were indicative that the effect of age and gender appears to be significant ($P < 0.05$) on all comparisons except males below 50 years and females above 50 years of age. Males above 50 had the significantly ($P < 0.05$) highest serum triglycerides than all other groups. Multiple comparisons of serum triglyceride level in hypertensive patients belonging to different age and gender showed that effect of age and gender appears to be significant ($P < 0.05$) on all comparisons except that males of either age had significantly ($P < 0.05$) higher triglycerides than females of below 50 years. Multiple comparisons of mean serum triglycerides in patients of myocardial infarction belonging to different age groups and gender showed a significant ($P < 0.05$) effect of age on serum triglyceride level of males. However, the effect of age on serum triglyceride level of females was observed as nonsignificant. Females of either age group had significantly lower serum triglyceride than males below 50 years of age. In the case of control group, multiple comparisons of serum triglyceride level indicated a nonsignificant effect of age on triglyceride level of males. Females above 50 years had significantly ($P < 0.05$) high serum triglyceride than all other control groups. Comparison of serum triglyceride level in different study groups with control subjects showed that all study groups except females below 50 years of age had significantly high serum triglycerides than control group.

Index

N.B. Rizvi and S.A. Nagra, *Minerals and Lipids Profiles in Cardiovascular Disorders in South Asia*, DOI 10.1007/978-3-642-34249-3,

O

R

S

9783642342486